3 1316 00388 2530

D0985276

TEARDOWNS

About the Author

Bryan Bergeron (Brookline, Mass.) is editor of *Nuts & Volts Magazine* and *Servo Magazine* and the author of several hundred articles, two dozen books, and several patents. He teaches in the HST Division of Harvard Medical School and MIT. His company, Archetype Technologies, Inc., develops intelligent systems for the military.

Bryan Bergeron

New York Chicago San Francisco Lisbon
London Madrid Mexico City Milan New Delhi
San Juan Seoul Singapore Sydney Toronto

The McGraw-Hill Companies

Library of Congress Cataloging-in-Publication Data

Bergeron, Bryan P.
Teardowns : learn how electronics work by taking them apart / Bryan Bergeron.
p. cm.
Includes bibliographical references and index.
ISBN-13: 978-0-07-171334-4 (alk. paper)
ISBN-10: 0-07-171334-4 (alk. paper)
1. Electronic apparatus and appliances—Maintenance and repair—Technique.
2. Electronic apparatus and appliances—Design and construction—Technique.
3. Electronic apparatus and appliances—Experiments. I. Title.
TK7870.2.B47 2010
621.381—dc22 2010020192

McGraw-Hill books are available at special quantity discounts to use as premiums and sales promotions, or for use in corporate training programs. To contact a representative, please e-mail us at bulksales@mcgraw-hill.com.

Teardowns: Learn How Electronics Work by Taking Them Apart

1234567890 DOC DOC 109876543210

ISBN 978-0-07-171334-4
MHID 0-07-171334-4

Sponsoring Editor
Roger Stewart

Editorial Supervisor
Jody McKenzie

Project Manager
Aloysius Raj, Newgen Imaging Systems Pvt Ltd, India

Acquisitions Coordinator
Joya Anthony

Copy Editor
Lisa Theobald

Proofreader
Paul Tyler

Indexer
Karin Arrigoni

Production Supervisor
Jean Bodeaux

Composition
Newgen Imaging Systems Pvt Ltd, India

Illustration
Newgen Imaging Systems Pvt Ltd, India

Art Director, Cover
Jeff Weeks

Cover Designer
Jeff Weeks

For Fred Marshall, my first mentor in the world of science.

Contents at a Glance

Contents

PART III For Musicians

Foreword

Most electronics books for novices and hobbyists either teach the basics or present a set of do-it-yourself projects. This book breaks the rules by reverting back to what most of us did long before we even read an electronics magazine or book. We took things apart to find out how they worked. And that's precisely what Bryan Bergeron, the editor of *Nuts & Volts* magazine, has so expertly done in this remarkably interesting, nicely illustrated, and fun-to-read book.

While *Teardowns* is appropriately subtitled *How Electronics Work by Taking Them Apart*, the book is much more than this. It begins with an introduction that summarizes the keys to working with basic electronic circuits in only 13 fast moving pages, each of which is packed with practical tips and advice based on the author's extensive experience with hands-on electronics. The book then jumps straight to a series of commercially available electronic products that the author opens, pries into, disassembles, and photographs one step at a time.

The text is written as if Dr. Bergeron, who is a highly experienced electronics practitioner, is speaking directly to the reader with a point-by-point commentary about each teardown, complete with clear explanations of the operation and function of every component. By the time the product is completely disassembled, the reader understands the design tricks, component selection, and packaging choices that enabled the product to reach the market.

The teardowns are divided into three sections. Electronic products commonly found around the home are disassembled first, including a smoke alarm, motion-activated light, digital scale, ultrasonic humidifier, and five others. A section for tinkerers includes an intriguing teardown of an analog volt-ohm meter that effortlessly teaches more about Ohm's law than most college textbooks. A closing section on electronic music takes the reader inside a Fender guitar, an effects pedal, and even a vacuum tube audio amplifier. Along the way, Bergeron suggests various improvements, modifications, and enhancements that can be added to many of the products. He also explores the operation of the various sensors in many of the products he investigates, including those that detect light, pressure, sound, and the vibration of a metal guitar string.

The high quality of the many close-up photographs and circuit diagrams greatly enhances the value of this book. Considerable time was obviously devoted to the photographs, which are crystal clear with excellent lighting and depth of field. Want to know if your solder connections are well done? Just compare yours with the examples of sloppy and good connections in the photos. Want to see the different layout methods manufacturers use to install surface-mount components and conventional parts equipped with pins and wire leads? Just browse through the crisp photos of exposed product circuit boards in the first section.

In the end, the author's top-down approach to explaining electronics provides a unique learning experience and a user-friendly reference for both novices and experienced circuit builders. The book is also a useful teaching tool for electronics instructors. Old timers will wonder why we didn't think of this terrific book idea before the prolific Bryan Bergeron arrived on the scene.

Forrest M. Mims III Geronimo Creek Observatory, Texas www.forrestmims.org

The author of 60 books and a long-time contributor to *Nature, Scientific American, Sky & Telescope, Popular Mechanics, IEEE Spectrum*, and many other magazines and journals, Forrest Mims is the most widely read electronics author in the world. His clients have included the National Geographic Society, the National Science Teachers Association, and NASA's Goddard Space Flight Center. In 1993, he was named a Laureate in the Rolex Awards for Enterprise.

Forrest M. Mims III
Geronimo Creek Observatory, Texas
www.forrestmims.org

Acknowledgments

Every book starts as a spark that sets matter into motion. Roger Stewart, editorial director at McGraw-Hill Professional, provided the spark for this book. Special thanks to Joya Anthony, acquisitions coordinator, for her energetic, day-to-day support throughout the publishing process.

Introduction

Teardowns: Learn How Electronics Work by Taking Them Apart takes a unique, fun approach to learning about electronics by first guiding you through the disassembly process of popular electronic devices and then discussing the uncovered components and circuitry. That old, broken amplifier or electric guitar in your garage can be transformed from a dust collector into the perfect basis for tutorials on learning the principles of electronics. As such, you can follow the projects discussed in this book as the basis for your own teardowns, even if you are on a restricted budget. Moreover, *Teardowns* is designed as a stand-alone reference, so that you can follow and learn from the photographs and descriptions of teardowns without having access to a shop, tools, or specific devices.

As each device is disassembled, the interaction and function of the components are discussed, providing an immediate working context for otherwise abstract concepts. Moreover, this is much more than a "way things work" book, because the teardown process illustrates the physical nature of device construction and provisions for mechanical stability, heat transfer, radio frequency interference prevention, safety, and design for serviceability—information you won't find on a schematic or specifications sheet. In addition, when appropriate, an improvement or modification is made to the device and the results are discussed. In this way, that old amp rescued from the garage serves as a vehicle for learning and enjoys a second life.

The inspiration for this book stems in part from human nature and in part from my early experiences in electronics. You can probably recall a time when a toy or appliance broke and you followed your natural curiosity and disassembled the device, sometimes discovering a fix for the problem, but always learning about how the device was constructed. In my experience, the texts and booklets on beginning electronics provide the theoretical basis for electronics, and necessarily use over-simplified, easily digested circuits, often without a discussion of practical design considerations.

In contrast with my readings of texts and magazine articles, my most memorable early encounters with electronics were at the workbench, working under the tutelage of an experienced technician. I had informal lessons in diagnostics, around whatever piece of equipment he happened to be repairing. But among the more enjoyable experiences was the teardown. We'd start with a ridiculously inexpensive piece of nonfunctional equipment from a local flea market and either repair it or disassemble it for components and spare parts.

I learned to look for hidden treasures. For example, Tektronix used to include a length of silver solder hidden away in its oscilloscopes for repairs. I also learned what components tended to fail first in different equipment. Through repeated quizzing from my mentor, I learned to differentiate between various types of capacitors, inductors, and solid-state components.

The problem with unguided teardowns is that, without knowledgeable guidance, it's easy for a novice to miss subtle design considerations, such as the mechanical stability of electromagnetic interference (EMI) shielding, component placement that facilitates convection cooling, or the usefulness of knowing the pinouts of a standard transistor package. By providing virtual handholding to virtual teardowns, this book serves as a necessary adjunct to the numerous theory-based books and web sites.

Assumptions

In developing this book, I've made a few assumptions about you, the reader. If you're strictly reading along with the teardowns, I assume that you

- have read at least one introductory book on electronics.
- can recognize basic active and passive components.
- have some experience building basic circuits, either from a kit or a web/magazine article.
- have a natural and insatiable inquisitiveness.

In addition, if you plan to use this book as a field guide of sorts to real teardowns, I assume that, in addition to the above, you

- have access to and know how to use the basic tools of the trade: screwdrivers, cutting and needle-nose pliers, soldering iron, desoldering tools, and a multimeter.
- are aware of basic electrical safety issues.
- have and use eye protection. Odds are, if you get shocked by a charged capacitor or cut yourself with a screwdriver, you'll live and soon forget about it. However, if you happen to launch a piece of wire into your eyes with diagonal cutters, your vision could be permanently degraded—at least until bionic implants are available. Protect your eyes.

Goals

I have done my best to insure that, after reading this book, you will

- have a better understanding of the electronics underlying common electronics devices. The teardowns should give you mental anchor points for an otherwise abstract theory of electronics.

- better understand the connection between a physical device and its corresponding schematic. I'll offer a schematic for each device disassembled.
- be able to tear down and understand electronic devices not covered in the book. I'd like you to be able to tear down any device without fear of personal injury or of permanently making the device inoperable.
- better understand the practical considerations that go into electronic product design from both functional and manufacturing perspectives. It's one thing to design a circuit for a textbook exercise, and another to consider how it might be stressed in a high-temperature environment, when subject to vibration and other physical abuse, and used with an imperfect power source.

Organization

Teardowns is organized in three main parts, based on the types of electronic devices discussed, and two appendices. As you'll see, I don't treat components equally from one teardown to the next, but instead focus on what makes a particular device unique. For example, resistors and Ohm's law are merely mentioned in the teardown of an ultrasonic measuring device, but they take center stage in the teardown of an analog Volt-Ohm-Meter (VOM).

Part I: Around the Home

This section targets electronics you're likely to have in and around your home or that you might find at a garage sale. The components and topics covered in these nine chapters are outlined here.

Chapter 1: Dual-sensor Smoke Alarm

This chapter describes the teardown of a residential smoke alarm that features both photoelectric and ionization sensors. The following key components and topics are covered in this chapter:

- Ceramic disc capacitors
- Decibels
- Electrolytic capacitors
- Epoxy-dipped capacitors
- Infrared LEDs
- Ionization sensors
- Photodiodes
- Photoelectric sensors
- Piezoelectric transducers
- Radiation
- Resistors
- Silicon diodes
- Transistors
- Zener diodes

Chapter 2: Motion-activated LED Light

The focus of this teardown is a passive infrared motion-activated LED light. The following key components and topics are covered in this chapter:

- Ambient light sensors
- Battery amp-hours
- CDS (cadmium sulfide) sensors
- Circuit boards
- Electrolytic capacitors
- Fixed-voltage regulators
- Fresnel lens
- LED current limiting resistors
- Passive infrared (PIR) sensors
- SMT transistors

Chapter 3: Digital Bathroom Scale

The typical digital bathroom scale packs a lot of technology into a small, affordable package. This chapter describes the teardown of a popular digital bathroom scale. The following key topics and components are discussed:

- EEPROMs
- Gauge Factor (GF)
- LCD
- Microcontrollers
- Piezoelectric transducers
- SMT components
- Strain gauges
- Transistors
- Wheatstone bridge
- Zebra connectors

Chapter 4: Surge Protective Devices

This chapter features two teardowns of a common surge suppressor strip and a power conditioner sold to audiophiles and musicians. The following topics and components are covered in this chapter:

- Air gap insulation
- Avalanche diodes
- Bleeder resistors
- Bridge rectifiers
- Ceramic disc capacitors
- Electrical isolation
- Electrolytic capacitors
- Electromagnetic interference (EMI)
- Ferrite inductor
- Filter circuits
- Flame-proof resistors
- Illuminated rocker switches
- Joule rating
- Linear filtering
- Metal oxide varistors (MOVs)
- Metalized polyester film capacitors
- Mylar capacitors
- NEMA plugs and jacks
- Neon bulbs
- Relays
- Response time
- Silicon diodes
- Slow-blow fuses
- Thermal circuit breakers and fuses
- Thermal MOVs (TMOVs)
- Toroidal inductors
- Transformers
- Transient voltage suppression (TVS) diodes
- Tubular ferrite chokes
- UL 1449 rating
- Visible light LED
- Voltage regulators

Chapter 5: Electronic Pedometer

Many dedicated handheld electronic devices, from GPS units to cameras, have been consumed by the smartphone. This chapter describes the teardown of one of the last holdouts, the stand-alone electronic pedometer. The following topics are covered in the teardown:

- Circuit board layout
- Elastomeric buttons
- LCDs
- Mechanical pendulum sensor
- MEMS (micro electro-mechanical machine system) accelerometers
- Microcontrollers
- Piezoelectric pendulum sensors
- Potentiometers
- SMT components
- Tuning fork crystal oscillator
- Zebra connectors

Chapter 6: Compact Fluorescent Lamp

The compact fluorescent (CF) lamp is an evolutionary paradox. Although more energy-efficient than a small length of tungsten wire, the complex device violates most of the laws of basic engineering. The key concepts and components covered in this teardown include the following:

- Capacitive reactance calculations
- Color rendition index (CRI)
- Color temperature
- Diode for alternating current (DIAC)
- Diode peak inverse voltage (PIV) rating
- Electrolytic capacitors
- Electromagnetic interference (EMI)
- Ferrite core inductors
- Flame-proof resistors
- Fusible metal film resistors
- Heat sinks
- High-voltage measurement
- Lumens
- Mercury toxicity
- Power factor
- Silicon diodes
- Silver mica capacitors
- Toroidal transformers
- Total harmonic distortion (THD)
- UV radiation
- Voltage dividers
- Voltage doublers

Chapter 7: Ultrasonic Humidifier

This teardown unveils the hidden complexity of a popular ultrasonic humidifier. Key issues and components discussed include the following:

- Bicolor LEDs
- Bridge rectifiers
- Brushless DC (BLDC) motors
- Capacitive reactance calculations
- Ceramic disc capacitors
- Closed loop control systems
- Diode PIV rating
- Electrolytic capacitors
- Epoxy-dipped capacitors
- Fast-blow fuses
- Ferrite core inductors
- Hall effect sensors
- Heat sinks
- Inductive reactance calculations

- Iron laminate power transformers
- LC resonant frequency
- LED current limiting resistors
- Linear power supply
- Magnetic reed switches
- Metalized polypropylene film capacitors
- Motor controller IC
- Mylar film capacitors
- Open loop control systems
- Open-air inductors
- Piezoelectric transducers
- Potentiometers
- Response time
- RF oscillator
- Rocker switch
- Silicon power diodes
- Switching diodes
- Transistor current gain
- Transistor switches

Chapter 8: Digital Hygro Thermometer

This teardown introduces basic analog sensor circuits. Key component and topics discussed include the following:

- Analog sensors
- Circuit board layout
- Humidity sensors
- LCDs
- Measuring relative humidity
- Microcontrollers
- RC time constants
- SMT components
- Temperature coefficients
- Thermistors
- Tuning fork crystal oscillator

Chapter 9: Stereo Power Amplifier

This teardown features a popular, inexpensive stereo power amplifier with components and circuits you'll likely to find in any modern amplifier. The following key components and topics are discussed:

- ASICs
- Audio taper potentiometers
- Bridge rectifiers
- Bridging amplifiers
- Bypass capacitors
- Center-tapped transformers
- Chokes
- Class AB stereo audio amplifiers
- Decibels
- °C/W ratings
- DPDT switches
- Dual-polarity power supplies
- Frequency response
- Heat sinks
- Impedance matching
- Linear power supplies
- Metalized polyester film capacitors
- Mica insulators
- Mylar film capacitors
- Operational amplifiers
- Precision metal film resistors
- Relays
- Signal diodes
- Signal-to-noise (SNR) ratio
- Silicon power diodes
- Silver mica capacitors
- Slow-blow fuses

- Stacked potentiometers
- Switching diodes
- Tantalum capacitors
- Thermal management
- Toroidal power transformers
- Total harmonic distortion plus noise (THD+N)
- Visible-light LEDs
- Zener diode shunt voltage regulators

Part II: For Tinkerers

This section focuses on two electronic instruments you're likely to have (or want) on your workbench or toolbox: an analog VOM and a laser-guided ultrasonic distance measurer.

Chapter 10: Analog Volt-Ohm-Meter

This chapter reveals the circuitry and components found in a typical VOM. The following topics and components are discussed:

- Battery internal resistance
- D'Arsonval galvanometers
- Decibel measurement
- Diode forward voltage drop
- Diode PIV
- Electromagnetic buzzers
- Fast-acting fuses
- Full-scale accuracy
- Germanium diodes
- Linear potentiometer
- MOVs
- Meter sensitivity
- Nichrome wire resistors
- Precision metal film resistors
- Resolution, precision, and accuracy
- Rotary switch
- Schottky diodes
- Series and shunt resistors
- SMT components
- SMT resistor voltage rating
- Sound pressure calculations

Chapter 11: Laser-guided Sonic Distance Measurer

This chapter explores ultrasonic distance measurers and the physics of ultrasound. Key topics and components covered in this teardown include the following:

- Analog sensors
- Battery power
- Crystal ageing
- Elastomeric switches
- Electrolytic capacitors
- Fixed-voltage regulators
- Frequency response
- Heat sinks
- LC circuit resonant frequency
- LCDs
- Linear taper potentiometers
- Microcontrollers
- Operational amplifiers
- Resonant frequency calculations
- Signal-to-noise (SNR) ratio
- SMT components
- Sound propagation
- Step-up transformers
- Temperature coefficient
- Thermistors
- Ultrasonic transducers
- Variable voltage regulators
- Visible light laser
- Zener diode voltage clamp

Part III: For Musicians

The teardowns in this section, for your inner rock star, include an electric guitar, a tube-type guitar amplifier, and an effects pedal. The components and topics covered in these three chapters are outlined next.

Chapter 12: Electric Guitar

This chapter describes the teardown of a Fender American Deluxe Telecaster, including a detailed teardown of a magnetic pickup. Topics and components in this teardown include the following:

- Active and passive magnetic pickups
- Alnico magnets
- Audio jacks and switches
- Audio taper potentiometers
- Bypass capacitors
- Cable capacitance
- Capacitive reactance calculations
- Ceramic disc capacitors
- EMI shielding
- Frequency response
- Humbucking magnetic pickups
- Identifying magnet polarity
- LC circuit resonant frequency
- Measuring magnetic field strength
- MIDI
- Noise reduction
- Piezoelectric pickups
- RC filters
- Star ground configuration
- Wire insulation

Chapter 13: Effects Pedal

In this chapter, I'll tear down a popular distortion pedal for the electric guitar. The following key components and topics are discussed:

- Amplifier design
- Audio jacks and switches
- Audio taper potentiometers
- Ceramic disc capacitors
- Clipping
- Electrolytic capacitors
- Germanium diodes
- Intentional audio distortion
- Mylar film capacitors
- Operational amplifiers
- Silicon diodes
- Visible light LED

Chapter 14: Vacuum Tube Guitar Amplifier

The focus of this teardown is an all-tube 5W combo amplifier that's a favorite in the modding community because of its affordability and ease of modification. Key topics and components discussed in this teardown include the following:

- Audio output transformers
- Audio taper potentiometers
- Bleeder resistors
- Bridge rectifiers
- Bypass capacitors
- Center-tapped transformers
- Ceramic disc capacitors
- Chokes

- Class A audio amplifiers
- Electrolytic capacitors
- EMI shielding
- Enclosure acoustics
- Flameproof resistors
- Frequency response
- High-voltage power supply design
- High-voltage probe design
- Impedance matching
- Laminate core power transformers
- Linear power supplies
- Metalized polypropylene film capacitors
- Mylar film capacitors
- Silicon power diodes
- Slow-blow fuses
- Speaker design
- Thermal management
- Vacuum tubes

Appendixes

The appendixes include references to help you identify component values and learn more about specific topics introduced in the teardowns. Appendix A provides common component markings used on leaded and SMT capacitors, resistors, and inductors. Appendix B lists online and print resources for more information on components, design principles, and the topics discussed in each teardown.

If You Try This at Home

Sooner or later, you'll want to replicate at least one of the teardowns described in this book, and I encourage you to do so. Teardowns are fun, and there's no better way to learn the practical aspects of electronic device design and construction. However, if you're going to make the most of your educational experience, you'll have to exercise good judgment and take some commonsense precautions to avoid potentially dangerous situations.

Kill the Power

Before you start to disassemble an AC-powered device, verify that it's unplugged. I like to keep the AC plug on my workbench where I can see it, next to the device. Sometimes I'll wrap the plug with masking tape. Otherwise, in the heat of a teardown, you may hurriedly attempt to plug in your soldering iron, only to discover a few minutes later that you have a cold iron and a hot power supply.

Wear Eye Protection

Wear safety glasses. I like the polycarbonate DeWALT Reinforcer Safety Glasses with magnification (www.Dewalt.com) because they're affordable and lightweight, and the built-in magnifier is great for working with surface-mount components.

Cover Your Skin

A long-sleeve shirt and full-length pants are a good idea if you plan to work with a soldering iron. Solder can splash when you extract components from a circuit board. It's also a good idea to keep a pair of leather gloves handy and wear them when there's a risk of being pinched or punctured. Avoid synthetic gloves because a hot soldering iron tip cuts through them like butter.

Don't Inhale

Avoid breathing the fumes from molten solder. The rosin core used with solder is a powerful lung irritant. If you use a soldering iron during a teardown, make sure your work area is ventilated.

Wash Your Hands

Assume the circuit boards you handle are manufactured with lead solder, even if they state otherwise. Wash your hands after a teardown, especially before eating food or playing with children. Ingesting lead can lead to reduced IQ, anemia, attention problems, and seizures, among other problems. Look what lead poisoning did for the Roman Empire.

Beware of High-voltage and High-current Circuits

With the advent of low-voltage wall transformers and low-voltage components, serious electrical shocks from consumer electronics are relatively rare. Even so, I have seen a wedding band vaporized by a 12V, high-current circuit. Fortunately, the wearer didn't lose his finger, but he spent a fortune on reconstructive surgery.

If you tear down an amp or other device with high-voltage vacuum tubes, remember that high voltages can linger in the power circuitry for up to several days after the amplifier has been turned off and unplugged. My motto: When in doubt, short it out. With the amplifier unplugged, take an alligator clip lead, connect one end to the chassis ground, and carefully touch the other end to the positive (+) and negative (–) terminals on power supply filter capacitors. I usually clip a 10Ω, 5W resistor to the lead to limit the initial shorting current.

Don't Panic

You can stay safe by maintaining a sense of self-control. Resist the urge to take out the 5-pound sledge and flatten that circuit board that you've been trying to extract for a half hour. Take a break, come back, and look for the hidden screw (there's always at least one); then proceed thoughtfully. The goal, after all, is to learn and have fun. And electronics is best enjoyed when you're alive, alert, and intact.

Establish a Good Work Environment

For clarity, I photographed the majority of teardowns on a clean white tabletop. However, when working off-camera, I use an indestructible, butcher-block workbench with a hefty vise and power tools that I can use to crack open just about anything. For detailed circuit tracing and analysis, I have an illuminated, ventilated Formica work surface with most of my tools within easy reach (see Figure 1).

You don't need a special work surface for a teardown. Given an understanding spouse, parent, or roommate, a counter top or kitchen table is a fine place for a teardown, as long as you protect the surface from dings and scratches. Wherever you work, do your best to insure that it's well-lit, adequately ventilated, and immune to flying pieces of metal, glass and plastic, and the spray of molten solder.

Use the Appropriate Tools

From a safety perspective, use the tools most appropriate for the job. In other words, use a hex-head driver to remove a hex nut, as opposed to a pair of needle-nose pliers that are likely to slip and either pinch your hand or impale your foot. Not only will you save money on bandages, but your tools will last longer.

The hand tools you'll need for basic teardowns include Phillips and standard screwdrivers, needle-nose pliers, diagonal cutters, crescent wrench, and occasionally a soldering iron and solder wick braid. Eventually, you'll want to increase your arsenal of tools to include a pair of tweezers, a hex driver set, a power screwdriver, drill, and illuminated magnifying glass. A lightweight, desktop vise, such as one from PanaVise (www.panaviseonline.com), is great for working with small devices

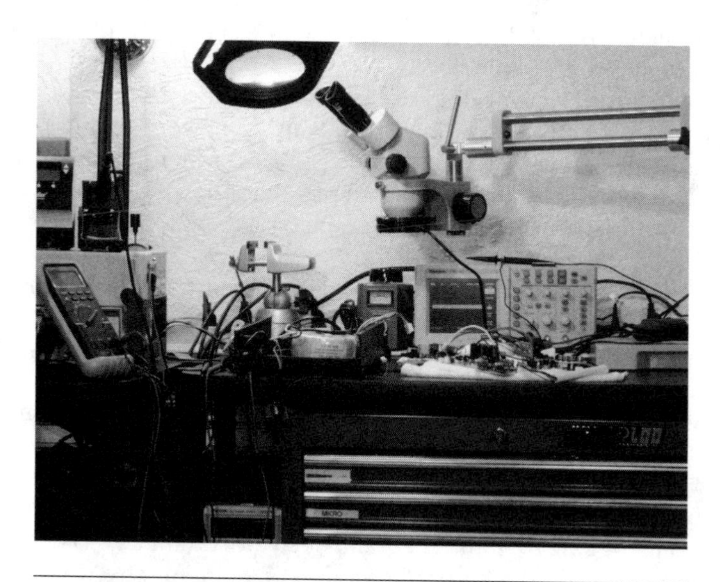

FIGURE 1 The author's work environment

and reworking circuit boards. Once you invest in a base, you can add a variety of heads, from circuit board holders to my favorite, the extra-wide opening head.

In the nice-but-not-necessary category are a variety of power tools. Depending on the task at hand, I choose between a cordless Dremel, a massive Milwaukee 1/2-inch drill, and Proxxon tabletop drill/milling machine. The ultra-precise Proxxon, an eBay find, is my favorite for reworking circuit boards. I also use a basic Weller WD1 soldering station for leaded components and a Weller WRS1002 hot air repair system that I picked up at a flea market for working with surface-mount components.

If you're tearing down a kitchen appliance, you'll want tough tools made for heavy gauge work. Conversely, if you're harvesting a delicate circuit and surface-mount components, you'll appreciate delicate, finely crafted tools. Diagonal cutters and pliers are rated for wire type and gauge. Learn the operating range of your tools. And don't forget the eye protection.

Use Instruments You Can Trust

The only instrument you need for basic teardowns is an inexpensive multimeter. You'll find the resistance function useful for tracing circuits and verifying component values, and the voltage function is indispensable for verifying that high-voltage capacitors are discharged.

If you work with high-voltage power supplies, consider purchasing a variable transformer. I admit that I worked with tube amps and high-voltage supplies for years without a Variac, but that's before the time of $20 audiophile capacitors and $250 power transformers. Besides, applying full voltage to dry electrolytic capacitors can make a real mess. I picked up new a 120V at 3A variable transformer with built-in voltmeter from All Electronics (www.allelectronics.com) for $60, and you can probably do much better on eBay.

If you can afford it, consider purchasing a Variac with built-in ammeter. A less-expensive alternative is to install a Kill A Watt EZ power monitor on the front end of any Variac you can get your hands on. The power monitor, available for about $35 from Amazon, provides a digital readout of voltage, current, and power factor, and you can unplug it and use it to assess the electrical equipment around your house.

Visualizing and assessing the polarity of magnetic fields can be important if you work with electric guitar pickups or electric motors. If you plan to work with magnets and electric motors, consider purchasing a magnet polarity tester and a magnetic field viewing paper. I use a $9 polarity tester from Stewart-MacDonald, a luthier supply shop (www.stewmac.com). The visualizing paper, available from Edmund Scientifics (www.scientificsonline.com) for about $3, is a laminated sheet of plastic containing iron filings suspended in oil. It's cleaner and more efficient than using iron filings.

In the nice-to-have category are an oscilloscope, a signal generator, and an automatic data logging system. The inexpensive Parallax USB Oscilloscope (www.Parallax.com) is a good starter oscilloscope. For more precise work and the ability to work away from a PC, I use a Tektronix TDS2022 color digital oscilloscope, a Fluke 87 DMM, and a Fluke 45 Dual Display Multi-Meter. For automated data logging, I use a USB-1208FS 12-bit data logger from Measurement Computing (www.mccdaq.com). The free software that accompanies the data logger is needlessly crippled in all the wrong places, but the hardware is bulletproof and a real time-saver.

Even if you haven't developed presbyopia, you'll find a magnified worklight handy. I use the three-diopter Electrix Halogen magnifier (www.electrixtask.com). It's bright, it provides a good working distance, and it's good for lighting tasks such as unsoldering leaded components. The downside is that the halogen bulb and anything close to it receives a thermal lashing. For detailed circuit tracing, I use a ProScope HR (www.bodelin.com/proscopehr) with the 100X illuminated head. While affordable, this handheld microscope tethers you to a laptop or desktop PC/Mac, and the working distance—the distance from the objective microscope lens to the subject—is only a few millimeters.

For serious surface mount work and identifying otherwise illegible component markings, I use a Luxo binocular microscope with dimmable LED ring light (www.luxo.com). It's expensive, but the working distance is about 5 inches—plenty of room for fingers, tweezers, and a soldering iron or hot air wand. This is definitely in the eBay, used equipment category.

Jotting down schematics in a notebook as you work is fine, but you'll eventually want to transfer schematics to something more legible. I've had good luck with ExpressSCH (www.expresspcb.com), which is free and easy to use. It's one of a dozen low-cost or free versions of schematic authoring tools on the Web.

Keep It Simple

If you're just starting out, keep it simple. Safety glasses, pliers, wrench, diagonal cutters, screwdrivers, and maybe a multimeter with the ability to measure capacitance and inductance are a reasonable arsenal of tools. You want to focus on the teardown, and not on how to operate a new battery of instruments.

The list of instruments in the preceding section may seem overwhelming at first—and it should. It represents my years of scouring eBay and local flea markets for deals on used but functional equipment. As you progress to designing and building your own circuits, you can gradually add the tools and instruments that suit your particular interests and budget.

Disclaimer

I've done my best to represent the devices and components fairly and accurately in the teardowns. As you'll note, many of the schematics are simplified, and all are based on my analysis. My objective is to highlight specific aspects of a circuit, not to provide an unofficial technical manual for the equipment. If you need a complete, official schematic, contact the manufacturer.

I independently selected and purchased the devices used in the teardowns, usually on Amazon or eBay. In fact, I purchased two and sometimes three of every device discussed in this book. I used one for a destructive, component-level teardown to determine, for example, the turns ratio of a transformer. The second device was for photographing and bench measurements. The occasional third device was for testing and verifying the operation of the second unit.

I don't have a financial interest in any of the manufacturers of the devices discussed in this book. I also don't have anything against any of the manufacturers. When something in a teardown disagrees with marketing material or specifications, I'll point it out. One thing that you'll learn from teardowns is that when the fine print on the packaging states "Specifications subject to change without notice," you're likely to find a few surprises in the teardown.

Web Site

The illustrations of teardowns in this book are necessarily limited to modest-sized grayscale images. To enhance your vicarious enjoyment, I've made additional media freely available at www.mhprofessional.com/computingdownload.

PART I
Around the Home

Chapter 1

Dual-sensor Smoke Alarm

Smoke alarms, such as the Kidde Pi9000 shown in Figure 1-1, are such commodity items that you probably rarely take notice of the white, saucer-shaped appliances that adorn the ceilings of homes, offices, and classrooms unless they're acting up. And if you've seen and heard one, you've seen and heard them all. Most sport a blinking red LED (light-emitting diode) to indicate the status of the battery or other power source. A test button lets you check out the audible alarm, and an obnoxious chirp indicates that it's time to change the battery. Your home unit likely has a hush button to kill the audible alarm when you accidentally trigger the unit by frying something a little too long on the stove.

FIGURE 1-1 Kidde Pi9000 smoke alarm

Reflecting this external homogeneity, little variability is found under the hoods of most models. The major differences between the less expensive and advanced commercial and residential units are connectivity and power. Wired or wireless connectivity, a central power source, and triggers for an autodialer for the fire department can double or triple the cost of a system. However, smoke alarms are the same where it matters most—in the sensor technology they use to detect smoke. And smoke inhalation is responsible for more deaths than is direct contact with fire.

The first technology is an *ionization-based sensor*, which is best for detecting the minute particulate by-products of burning liquids and other clean burning fires. The second technology is a *photoelectric sensor*, which responds best to the large particulate by-products from burning cigarettes, bedding, carpet, and other dirty, smoldering fires that produce optically dense smoke. Each technology has its strong points, and no single sensor is best for all situations. Many smoke alarms leverage the advantages of both technologies by offering dual sensors: an ionization sensor for clean fires and a photoelectric sensor for smoldering fires.

In this chapter, I'll tear down a dual-sensor Kidde Pi9000 smoke alarm. The basic sensors and, in many cases, the underlying circuitry and components are similar to units marketed by First Alert, Sentry, and many of the generic brands. In addition to exploring the overall design of this $20 system, I'll walk you through teardowns of the ionizing and photoelectric sensors. Because this is your first teardown, I'll discuss the circuits and components at a relatively high level. I'll dig deeper into the technology in the following teardowns.

Highlights

This is a quick teardown for the basic unit; the sensors are a little more involved and even potentially dangerous. The outer shell of the smoke alarm can be disassembled with a standard screwdriver. Once you're inside, the circuit board simply snaps out. The ionization sensor, the smoke sensor that uses ionization as the basis for its operation, is notable because it contains the synthetic element Americium-241, a potentially lethal alpha particle emitter with a half-life of about 430 years.

During the teardown, note the following:

- **The multiuse circuit board** Kidde apparently uses one board for several of its smoke alarms to cut down on inventory costs.
- **Component count relative to functionality** The Pi9000 is built around two chips that are designed for smoke alarms, which minimizes overall component count.
- **Construction of the two sensors, especially how they facilitate airflow** Particles have to make their way through the plastic outer shell of the alarm housing and then through the narrow slits of the sensors to be detected.

Specifications

Kidde doesn't provide typical technical specifications on its packaging, such as the minimum particulate count that will trip the alarm. The only information that comes close to a technical specification is the loudness of the alarm—85 decibels (dB)—but the number is meaningless because distance isn't specified. Information from the manufacturer of the piezoelectric transducer (which converts electrical pulses to mechanical vibrations—for the alarm sound) suggests that this distance is about 10 feet.

Despite the lack of specifications, the package does provide a feature list and a warning:

- **Dual sensors** An ionization sensor for fast, flaming fires and a photoelectric sensor for slow, smoldering fires
- **Hush button** To silence nuisance alarms
- **Test button and LED** For visual verification of battery and alarm operation
- **Warning chirp** To indicate that the battery needs to be replaced
- **Radiation notice** The product contains 0.9 microcurie (µCi) of Americium-241

Operation

Once the battery is installed, alarm operation is automatic. In normal conditions, the red LED flashes every 30 to 40 seconds to indicate that the unit is operating correctly. When a significant level of smoke is present, the alarm sounds, signaling that it's time to get moving.

Manual operation is limited to pushing and holding the test button and bracing yourself for the blast of audio. If the alarm trips because you're burning something on the stove, you can push the hush button to replace the continuous alarm with a less annoying chirp. However, if you're still smoking up the kitchen after 7 minutes, the alarm logic will override the hush command, and you'll have to bear with a continuous alarm.

Teardown

The teardown of the basic unit is illustrated in Figure 1-2 and discussed in the sections that follow. (I'll discuss the sensors later in the "Components" section.) If you plan to follow along and examine a physical smoke alarm, give yourself about 30 minutes to complete the teardown.

The ionization sensor contains a small amount of Americium, which is extremely dangerous if you inhale or ingest it. If you do plan to replicate this teardown at home, your safest course of action is to skip the teardown of the ionization-based sensor. It's also mandatory that you wash your hands after the teardown.

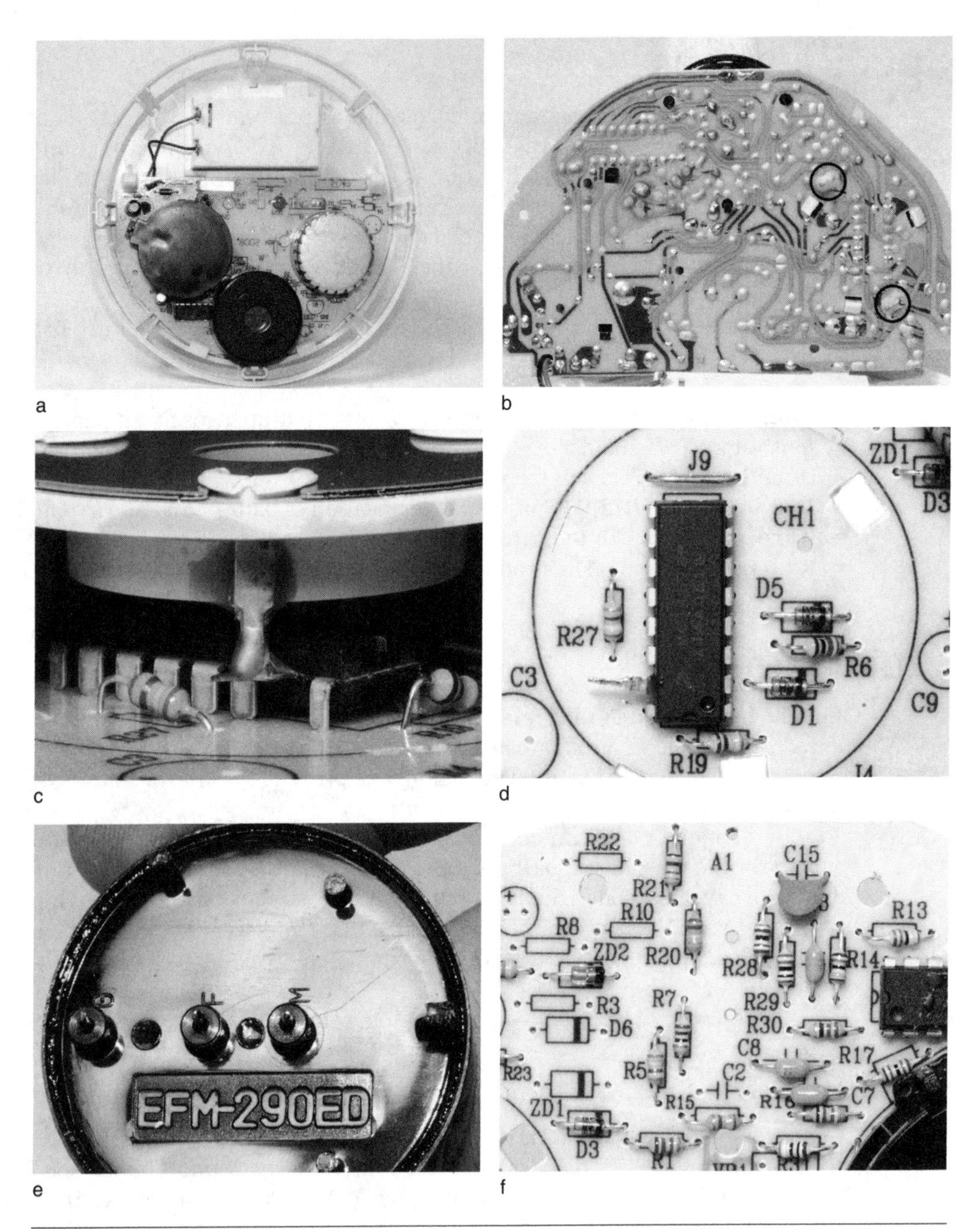

FIGURE 1-2 Teardown sequence

Americium is an alpha particle source. Alpha particles, two protons and two neutrons, are dense and relatively slow and travel perhaps 3 feet in open air. They are easily stopped by a sheet of paper, clothing, and even thick skin. The greatest potential for harm results from inhaling or ingesting the radioactive element. The linings of your lungs and intestines are only a few cells thick, much thinner than a sheet of paper or your hide, and are highly susceptible to radiation damage. So don't try to cut, file, sand, or remove the pencil eraser–sized Americium disc in the ionization-based sensor.

Tools and Instruments

This teardown can be performed with a small standard screwdriver, an old pair of diagonal cutters, a soldering iron with the usual soldering and desoldering accessories, a multimeter, and a magnifier. A Geiger counter or other alpha-radiation detector is required to assess the level of alpha radiation emitted by the ionization source.

If you try this teardown at home, do not crack open the ionization sensor unless you have a means of monitoring the radiation level.

Step by Step

Accessing the single circuit board in the smoke alarm involves little more than removing a plastic cover and unsnapping the circuit board from the plastic base.

Step 1

Crack the case. With the unit face down and the battery compartment empty, use a standard screwdriver blade to disengage the four equally spaced plastic hooks that hold the two halves of the clamshell case together. Remove the front half—the side normally facing down from the ceiling—and set it aside. Orient the bottom half of the alarm so that the 9V battery compartment is away from you, as shown in Figure 1-2a.

Step 2

Identify the sensors. The three prominent cylindrical components, from left to right, are the photoelectric sensor, the piezoelectric transducer, and the steel-encased ionization sensor. I'm tempted to refer to the piezoelectric transducer as the *buzzer* for simplicity, but as you'll see later, it requires an external circuit to create the alarm tone.

Note that each of these components is open to the air. The two sensors feature vertical slits to take in particulate by-products from fire while excluding insects, dust, and, in the case of the photoelectric sensor, light. The cover of the piezoelectric transducer has a large hole in its center to enable the internal piezoelectric disc to interface directly with the air to create ear-piercing sound waves. Locate and identify the 16-pin MC146012ED IC (integrated circuit), just to the left of the piezoelectric transducer and adjacent to the photoelectric sensor.

Step 3 (Optional and Potentially Hazardous)

Remove the outer steel can of the ionization sensor. Remember that this is an optional step that can potentially expose you to alpha radiation.

From a personal liability perspective, I advise against removing this sensor can unless you have supervision from someone knowledgeable about radiation. At a minimum, have a Geiger counter by your side so that you can determine if and when you're being exposed to radiation. Without a counter, there's no way to know if the outside of the sensor is contaminated, for example.

If you're equipped to follow along at this point, put on a pair of latex gloves. The inexpensive surgical variety is fine, but you'll get extra protection by wearing the thicker gloves sold for dishwashing. Next, locate the two metal tabs from the slotted steel ionization sensor cover on the underside of the board, circled in Figure 1-2b. Unsolder the tabs and remove the can to reveal the plastic and steel ionization chamber. Note the MC14578P IC, tucked beneath the chamber, as shown in Figure 1-2c.

The IC is a standard 16-pin DIP (dual inline package) with pin 1 marked by the depression in the top of the package. In Figure 1-2c, pin 1 is in the lower-left corner of the IC, pin 8 is in the upper-right corner, pin 9 is in the upper-left corner, and pin 16 is in the lower-left corner. Don't forget that the pin numbering system is clockwise if you're looking at the underside of the component.

Step 4 (Optional and Potentially Hazardous)

Extract the ionization chamber from the circuit board. Unsolder the connection between the ionization chamber and straightened pin 15 of the MC14578P beneath the chamber, visible in Figure 1-2c. Unsolder the tab from the top plate of the ionization chamber where it connects to the underside of the circuit board, near pin 7 of the IC. Locate the three white plastic tabs on the underside of the board, adjacent to where the two metal tabs were removed from the board. Cut the plastic tabs with a pair of diagonal cutters or push them through the board with the tip of your screwdriver.

Pull the ionization chamber straight up to for a full view of the MC14578P and a few external components, as shown in Figure 1-2d. Put the chamber in a plastic freezer bag and place the sealed bag in a metal drawer or other container. The plastic bag should contain the alpha radiation, but the added protection provided by a metal container can't hurt.

Step 5

Temporarily remove the piezoelectric transducer from the circuit board. From the underside of the circuit board, unsolder the three leads of the transducer. Examine the bottom of the extracted transducer, shown in Figure 1-2e. Note the three terminals, marked *G*, *F*, and *M*. Examine the transducer with an ohmmeter while you have the component out of circuit. You should read several megohms (MΩ) between each terminal. You'll have to replace the transducer to test the logic and sensors, but for now set the transducer aside.

Step 6

Examine the components that were hidden under the piezoelectric transducer, shown in Figure 1-2f. Note the outlines of components that were not installed by the manufacturer. Now is a good time to trace the circuitry of the alarm.

Step 7

Extract the photoelectric sensor. Unsolder the two leads to an IR (infrared) LED at the 10 o'clock position of the sensor and two leads and a tab to an IR photodiode at the 5 o'clock position. You can't see the IR LED or IR photodiode at this point in the teardown because they're hidden by the case of the sensor. Note the C8050 NPN transistor and the silicon power diode external to the sensor at the 11 o'clock position.

Step 8

Replace the piezoelectric transducer. Solder the transducer back in place. Then take a break and wash your hands. Even if you didn't extract the ionization chamber, it's a good idea to wash your hands because of possible contamination with Americium.

Layout

As revealed during the teardown, the 9V battery is located in a separate compartment in the base. I found it easier to work with the circuit by unsoldering the connection to the compartment and using a bench supply for power.

The single-sided circuit board shows stencils for components not included in this product, including a transistor, several diodes, and a few capacitors, probably for the more expensive models with recorded voice alarms. Other than the unorthodox method of soldering the IC pin directly to the ionization sensor and the use of inexpensive spring switches, this is an unremarkable layout.

Components

The two 16-pin chips, the ionization and photoelectric sensors, and the piezoelectric transducer make up the bulk of the smoke alarm circuitry. Let's consider the ionization and photoelectric detection systems separately.

Ionization-based Sensor

The ionization-based sensor is an ionization chamber that requires an MC14578P or equivalent discrete components to perform as a smoke detector. The ionization chamber is constructed of two discs separated by about 1 centimeter of air. The top disc is perforated in the center to allow air and smoke into the chamber, as shown in Figure 1-3. The figure shows the perforated top of the chamber partially removed, revealing a glimpse of the bottom disc.

A pellet of Americium-241 is firmly embedded in the center of the bottom disc. Figure 1-4 shows the view from the top of the sensor. The pellet does not extend through the bottom disc, thereby containing the radiation to the area above the bottom disc. Because the top disc is perforated, ionization leaves the chamber but is stopped by the steel cover that was removed earlier in the teardown.

According to the package back, the pellet contains 0.9μCi of Americium-241. The curie (Ci) is a unit of radioactivity related to the number of decays per second:

$$1\text{Ci} = 3.7 \times 10^{10} \text{ decays/second}$$

$$0.9\mu\text{Ci} = 3.3 \times 10^{6} \text{ decays/second}$$

The number of decays per second, indicative of the number of alpha particles kicked out of the pellet per second, is key to this application, because the alpha particles keep the air within the ionization sensor ionized. And ionized air conducts, unless it's filled with smoke.

If you have access to a radiation detector, you can test the activity of the Americium within the ionization chamber, as shown in Figure 1-5. As you can see, a few centimeters from the source the activity is 1.56 thousand counts per

FIGURE 1-3 Ionization chamber with perforated top partially removed

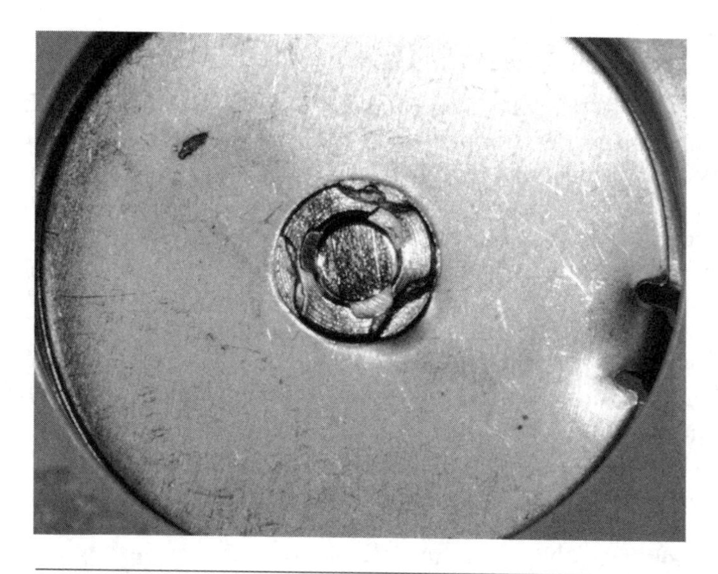

FIGURE 1-4 Americium-241 embedded in the bottom disc of the ionization chamber

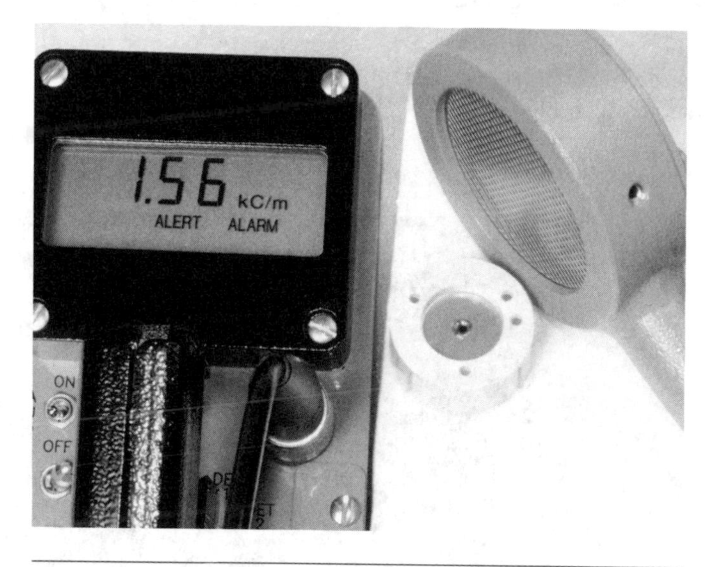

FIGURE 1-5 Radiation meter reading Americium alpha particle activity

minute (kC/m), or about 26 counts per second—enough radiation to cause problems if the Americium is handled carelessly.

If you handle the ionization chamber, make certain to wash your hands immediately afterward.

Ionization Sensor IC

The 16-pin MC14578P is designed specifically to interface with an ionization chamber as part of a smoke alarm. The IC is a CMOS (complementary metal oxide semiconductor) micro-power comparator that requires only one external component—a resistor. The comparator is configured so that minute changes in ionization current result in massive negative or positive swings in the comparator output, depending on the relative change in input of the comparator.

A key parameter of the ionization sensor IC is a standby current of only 10 microamps (μA) and a comparator input current of 1 picoamps (pA). In other words, the chip loafs along consuming only 10μA as long as the ionization current is stable. This ultra-low power consumption contributes to the longevity of 9V batteries in this home smoke alarm.

The chip can also supply 25 milliamps (mA) output current when triggered by a change in ionization current, which is more than enough to signal another device in the smoke alarm. For more detailed information on the chip, see the technical data document available from Freescale Semiconductor, Inc. (www.freescale.com).

Photoelectric Sensor

The photoelectric sensor consists of an IR photodiode and an IR LED mounted in a black, light-tight chamber with slits that are wide enough to allow air and smoke in. As shown in Figure 1-6, the IR LED is oriented at the 1 o'clock position, but the

FIGURE 1-6 Photoelectric sensor with cover removed

photodiode is at about the 2 o'clock position. An obstructing wall is positioned near the center of the sensor. Together, the positioning and obstruction wall ensure that, under normal conditions, no direct IR radiation from the LED strikes the photodiode. Light from the LED is absorbed by the corrugated black surface inside the photoelectric sensor.

However, when large particles from a smoldering flame enter the slits of the chamber, they reflect light from the LED, and this radiation is picked up by the photodiode. The more reflected radiation that hits the photodiode, the greater the photodiode current. At a predetermined level, the current trips the alarm and the piezoelectric transducer is activated.

To facilitate testing and experimenting with the sensor, I extracted the LED and IR photodiode from the black plastic case and soldered them in their normal configuration on the circuit board. Figure 1-7 shows the photodiode and IR LED remounted on the circuit board without the light-tight sensor shell.

Photoelectric Sensor IC

The heart of the photoelectric sensor is the photoelectric sensor IC, an MC146012ED, which is marketed by Freescale Semiconductor as a "Low Power CMOS Photoelectric Smoke Detector IC." The chip is designed for one thing: to sense reflected light from large smoke particles and generate an alarm by driving an external piezoelectric transducer. Because the ionization sensor chip does not

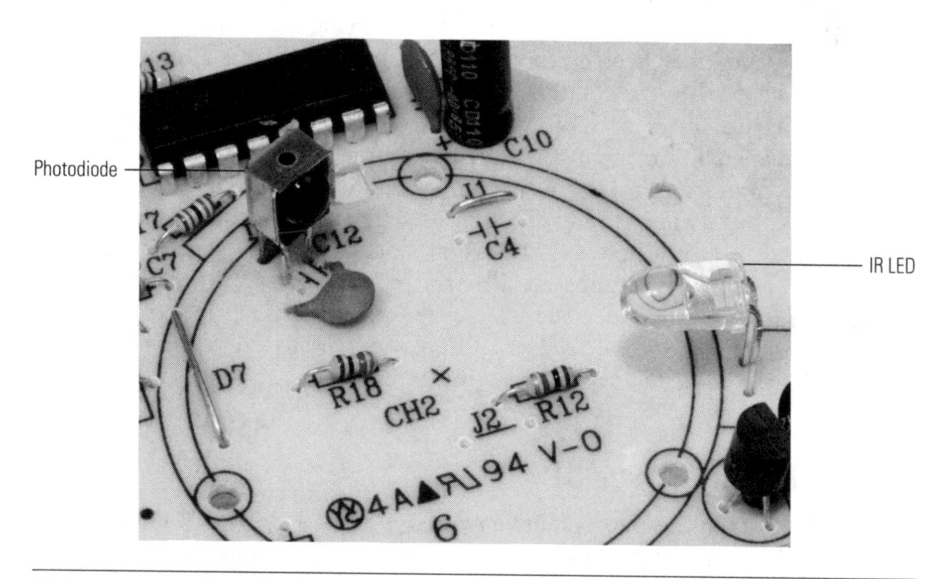

FIGURE 1-7 Photodiode and IR LED free of photoelectric sensor enclosure

have an onboard driver for the piezoelectric transducer, it has to piggyback on the photoelectric detector IC to sound the alarm.

The photoelectric sensor IC provides the logic for the smoke detector, from determining how long to remain silent after being hushed by the user, to when to flash the red LED, to how long to sound while the user is operating the test function. A handful of resistors and capacitors define the sensitivity of the chip. When hibernating, which is most of the time, the chip draws only 6.5µA.

Every 32 seconds, the chip awakens and sends a pulse to the C8050 NPN transistor and reads the photodiode. This gated IR LED pulsing and sensor reading not only minimizes power consumption, but it reduces the effect of noise from other light sources. This cycle continues until the photodiode detects IR light, at which time the chip activates its pulse/read circuitry every 2 seconds.

Piezoelectric Transducer

According to the manufacturer's product sheet, this piezoelectric transducer is capable of emitting a 3.5 kilohertz (kHz), 90dB tone at 12 inches when powered by a fully charged 9V battery.

Because 12 inches is a typical working distance on a bench and 90dB is enough to damage your hearing permanently, it's a good idea to tape over the sound orifice to reduce the output level. Even so, limit exposure to the audio blast, for your sanity if not your hearing.

According to my measurements, the transducer draws only 10mA at full output, which is amazing, given the ear-splitting level. As you discovered during the teardown, the three connections, G, F, and M, connect to dedicated pins on the photoelectric sensor IC. Pin G is the metal support electrode, pin M is the ceramic electrode, and pin F is the feedback electrode. Figure 1-8 shows the pin connections for the paper-thin transducer, partially extracted from the transducer case.

The practical significance of the transducer pin designations is to insure the proper interfacing with the modulation and driver circuitry within the photoelectric sensor IC. As with the two ICs, this is a specialized component specifically designed for use in alarm systems. Recall that this isn't a buzzer, in that no oscillator circuitry resides in the transducer body. However, the transducer is self-resonating at 3.5kHz when properly driven. However, if it's connected to a driver and excited at, say, 2.0 or 6.0kHz, the transducer will likely just sit there.

C8050 Transistor

The IR LED in the photoelectric sensor is pulsed by the C8050 general-purpose silicon NPN transistor that is commonly used as an audio output amplifier in portable radios. This 1 watt (W) transistor, which is packaged in a TO-92 case, is driven by the photoelectric smoke detector IC. Although the IC is probably

FIGURE 1-8 Piezoelectric transducer construction

capable of driving the IR LED directly, offloading the heavy lifting to the transistor increases the robustness of the circuit and probably extends the functionality of the IC at low battery voltages.

Diodes

This board contains power, switching, and zener diodes. A 1N003 silicon power diode placed across the battery leads where they enter the board insures correct battery polarity. If you accidentally reverse the 9V battery during installation, the diode will shunt the current, preventing the destruction of the two ICs, which are not reverse voltage protected. This isn't the most elegant use of a diode to insure proper battery polarity, in part because it may work only once, ending in the failure of the diode. However, it's efficient. The primary alternative is placing the diode in series with battery supply, but this drops the available voltage by the forward voltage drop of the diode to about 0.7V and wastes energy as heat.

The 1N4148 silicon fast-switching diodes are the most numerous diodes on the board. These glass-cased diodes can handle 200mA of forward current and 100PIV (peak inverse voltage).

The only zener diode on the board is in an unmarked glass package. A zener diode is designed to present a specific reverse breakdown voltage, typically much lower than a typical silicon power diode. The challenge of working with zener diodes is in limiting the reverse current with a series resistor so that the diode doesn't fail.

To determine the breakdown voltage of this zener, I removed the diode from the board and built a simple test circuit with a 4.5M series resistor. I hooked up the circuit to my bench power supply and slowly cranked up the reverse bias voltage across the resistor and zener diode while monitoring the voltage across the zener diode. With 5 to 10VDC (Voltage Direct Current) applied to the series circuit, the voltage across the zener diode varied between 2.5 and 2.8VDC. This test strongly suggests the diode is a 2.5V zener.

If you check the documentation for the ionization sensor chip, you'll see that a zener diode is specified to limit the voltage that can be applied to the negative comparator input pin. This limit is apparently 2.5VDC.

Capacitors

One electrolytic capacitor on the board, rated at 100 microfarad (µf) at 16VDC, is part of the transistor driver circuit. The remaining capacitors support the two ICs, as per the requirements listed on the datasheets. There include three ceramic disc capacitors and three axial lead epoxy-dipped capacitors, about the size of 1/8W resistors. Dipped capacitors are typically used instead of less expensive ceramic disc capacitors when specific capacitance values aren't available and when the circuit demands good temperature and humidity stability.

Switches

The switches in this smoke alarm are notable in that they are constructed of open spring steel that connects with a length of wire on the circuit board when compressed. Given that they'll probably be used a dozen times over the course of five years, someone made the economic decision to go with the cheapest possible switches. I can't think of a cheaper switch mechanism that works most of the time. These switches could be problematic if they become dusty, however.

How It Works

If you've followed the discussion of components to this point, you should have a good idea of the inner workings of the smoke alarm. As shown in the simplified schematic in Figure 1-9, a 9V battery supplies power and a silicon power diode (D1) protects against battery polarity reversal. The MC14578P chip (IC1) together with the ionization chamber forms the ionization sensor. A high-impedance operational amplifier (op amp) configured as a comparator within IC1 monitors current through the chamber.

In normal operation, alpha particles emitted by the Americium-241 ionize the air molecules within the ionization chamber. Electrons are knocked off of the atoms of gas in the chamber, forming equal numbers of positive and negative ions. Because one of the plates is held at a positive charge and the other at a negative charge, a small current flows as positive charged atoms flow toward the negative plate and negatively charged electrons flow toward the positive plate.

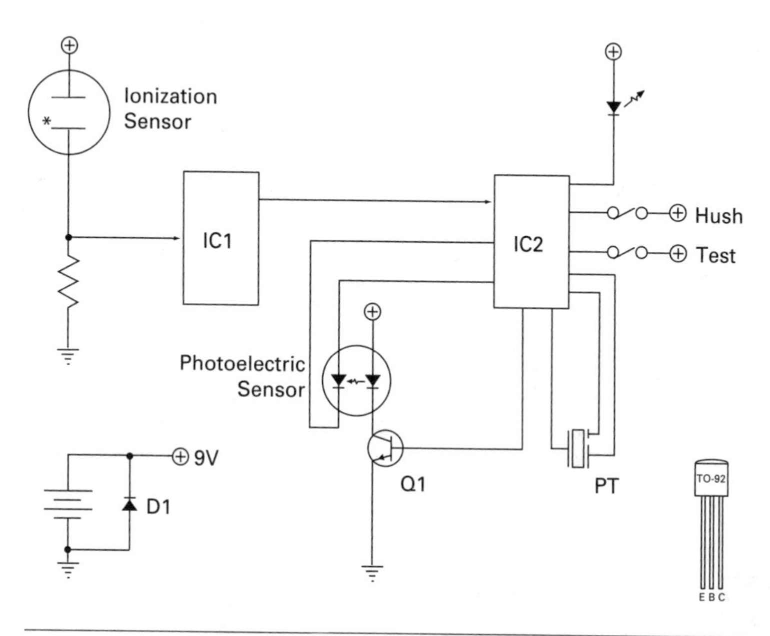

FIGURE 1-9 Simplified schematic

This ionization-induced current is stable until relatively massive smoke particles enter the cloud chamber. Smoke particles each absorb multiple ions, which reduces current flow. If the current flow falls below a predetermined level, an alarm is triggered. A limitation of this basic design is that the sensor current can be adversely affected by changes in humidity and atmospheric pressure, and this could affect the sensitivity of the alarm.

If smoke of sufficient density enters the chamber, the current decreases below a threshold established by the external resistors and capacitors connected to IC1. The chip sends an alarm condition to the MC146012ED photoelectric sensor chip (IC2).

When the air is free of smoke, IC2 periodically triggers the firing of the IR LED through the C8050 NPN transistor (Q1). If light is detected by the IR photodiode or if there is a trigger event from IC1, IC2 sends a 9V square wave to the piezoelectric transducer (PT), resulting in a loud alarm. IC2 also handles the logic of the hush and test buttons as well as the periodic flashing of the red LED. The data sheets for IC1 and IC2, which are readily available on the Web, define the signal levels and components needed for each pin and each function.

Note the outline of the transistor's TO-92 package in the schematic. It's a good idea to become familiar with this and other standard Transistor Outline (TO) packages because you can learn a lot from the package. For example, the standard TO-92 package defines the transistor leads as emitter, base, and collector, from left to right, facing the flat front of the transistor. The package size and composition also reveals a lot about the power rating of the transistor. For example, it's rare to find a transistor in a TO-92 package with a rating greater than 2W.

Mods

The Americium in the ionization sensor is an inexpensive, readily available source of radiation for testing radiation detectors, building a miniature cloud chamber, and performing other experiments using low-level radiation. However, unless you have the experience and equipment for monitoring radiation, it's best to avoid potential harm. Instead, let's focus on the IR photoelectric source and receiver.

External Alarms

If you download the datasheet for the IR photoelectric chip from Freestyle, you'll see that pin 7 is an input/output pin that can be used to activate escape lights, autodialers, and other electronic devices. You can experiment with this output to signal your PC through a serial or USB (Universal Serial Bus) port, or interface directly with a microprocessor.

I'm not advocating that you modify the smoke alarms in your home and then put them back into service. What I am suggesting is that you use a smoke alarm as an experimental platform for studying the underlying electronic circuitry. I'm sure that there's a law somewhere that imposes fines and prison time for anyone who tampers with a smoke detector, even in his or her own home. Besides, do you really want to entrust your life to an experimental circuit?

New Application Areas

The test setup shown in Figure 1-7 enables you to work with the LED and photodiode without the constraint of the light-tight container. You can also extend the leads of both components for remote sensing. For example, if you make each component waterproof, can you create a sensor to determine the turbidity of water? Could you use the detector circuit to create a pollen count detector?

Chapter 2

Motion-activated LED Light

Many electronic devices found in the home are interesting gadgets that briefly entertain and then become dust magnets. Other devices fulfill a real need, such as enabling you to navigate to the bathroom in the middle of the night without tripping over a toy or slipper. Of course, I'm referring to the motion-activated lights that turns on automatically when they detect your presence, such as the unit shown in Figure 2-1. The most useful of these lights employ battery power and an LED light source so that they can be placed anywhere they're needed, without regard to the availability of AC power.

Several sensor technologies can be used as a trigger mechanism for automatic lights, ranging from breaking a beam of light, to noise detection, to measuring

FIGURE 2-1 Fulcrum motion-activated LED light

fluctuations in a radiofrequency (RF) field; to detecting body heat, CO_2 concentration, or other products of respiration; to image recognition. Detecting body heat is a common, inexpensive sensor technology, in part because of the ready availability of passive infrared (PIR) detectors and dedicated support chips that minimize component count, manufacturing costs, and space requirements.

The focus of this teardown is the passive infrared Fulcrum motion-activated LED light shown in Figure 2-1. In addition to walking you through the circuitry, I'll suggest a few mods that will enable you to use the sensing circuitry of the battery-operated unit in additional applications.

Highlights

This teardown involves removing a few screws and extracting two circuit boards. The main board is populated by a variety of surface mount technology (SMT) and leaded components, including ambient light and PIR sensors; a dome-shaped Fresnel lens; a bank of white, high-intensity LEDs; a dedicated PIR integrated circuit; a voltage regulator; and an optoisolator. During the teardown, note the following:

- The quality of the solder joints on the circuit board
- SMT component markings
- Where leaded components are used instead of surface mount components
- Construction of the LED assembly

Specifications

Functionally, this is a relatively simple device. As such, it's not surprising that few specifications are provided by the manufacturer:

- The unit requires 3 AAA cells
- Light duration after triggering is 20 ±4 seconds
- Detection cone is 40 degrees at 10 feet
- The light consists of four white LEDs with an expected lifetime of 50,000 hours

We can deduce from the battery requirement that the device operates on 4.5VDC, assuming the AAA cells are wired in series. It's odd that the specifications include no mention of battery life, but we can measure the average current draw to determine the longevity of the cells under normal use. The detection angle and distance reflect the geometry of the optics and the sensitivity of the PIR sensor. LED lifetime is below average, but much better than equivalent incandescent lights. Overall, these specifications are typical for an inexpensive motion-activated light.

Operation

Once the unit is fitted with batteries and installed on a wall or other surface with double-sided adhesive tape, operation is automatic. If the ambient illumination level is low, the bank of LEDs is energized whenever the sensor detects a change in the level of ambient infrared (IR) radiation. A light sensor in the unit inhibits activation of the LEDs when the area of coverage is illuminated by daylight or an artificial light source. The LEDs remain active for about 20 seconds if no further change in radiation is detected. Otherwise, the LEDs stay on for as long as fluctuations exist in the level of ambient IR radiation.

The PIR sensor technology isn't foolproof, especially when it comes to false triggering. Any source of IR radiation can trigger the unit, from a dog or cat to a nearby heater, heat vents, or other heat source. Reflective surfaces near the sensor can redirect IR radiation from other areas and devices in a room. Fortunately, the worst-case scenario for false triggering is reduced battery life.

Teardown

You should be able to complete the teardown, illustrated in Figure 2-2, in about 10 minutes. If you're following along on your workbench with a battery-powered motion-activated light, you should encounter essentially the same sensor and control configuration. If you're working with an AC-powered unit, you may encounter an additional relay or other high-voltage switching component.

Tools and Instruments

The tools required for this teardown include a small Phillips-head screwdriver, an illuminated magnifier or strong magnifying glass, a multimeter, and forceps. If your circuit board was produced in the same plant as mine, you'll probably have to use the forceps to pull off mounds of excess glue to read component markings. Despite the glue and poor soldering work, with a little effort and a good magnifier, you should have no trouble tracing the circuitry.

Step by Step

There are no surprises in this teardown, other than having to remove gobs of hot melt glue from the main circuit board. Exercise caution when removing the optics so that you don't damage the mounting tabs, and avoid tugging on the fragile 28AWG (American Wire Gauge) wire connecting the battery holder to the main circuit board. Otherwise, the parts involved in the teardown are fairly indestructible.

If you're unfamiliar with the AWG system of classifying wire, now is a good time to get a feel for wire gauge. The smaller the AWG number, the thicker the wire.

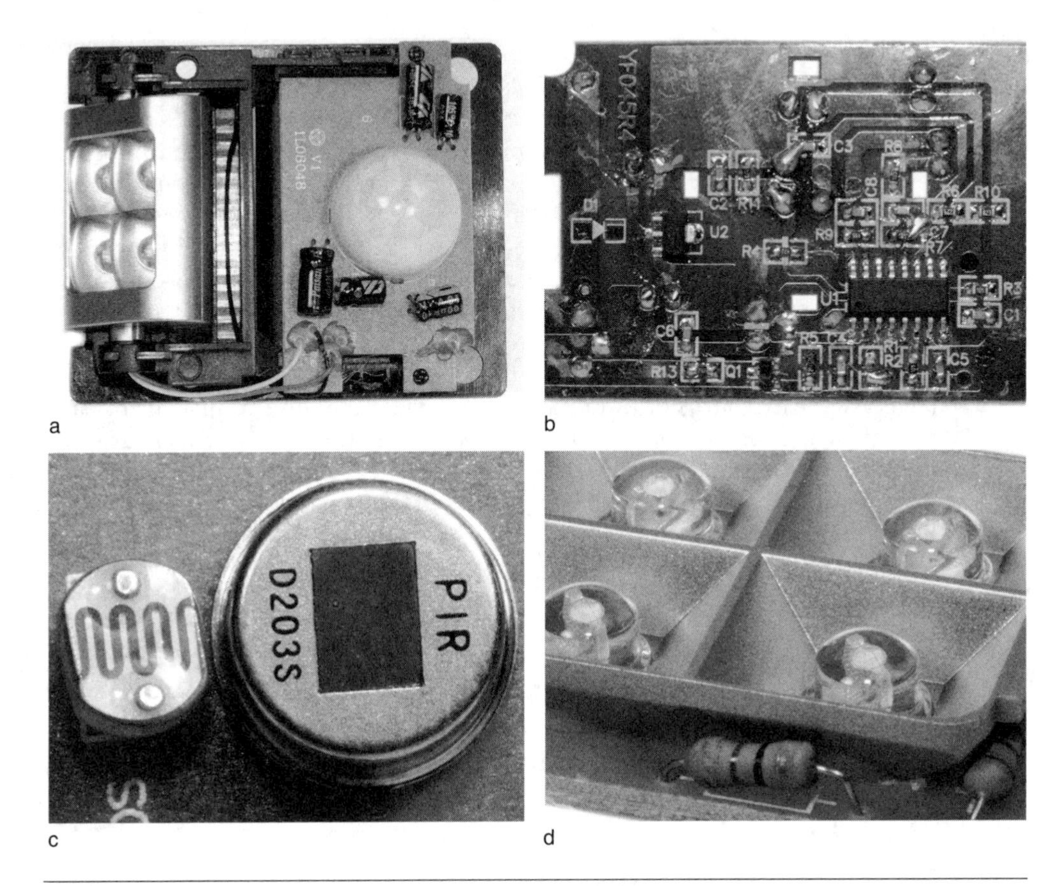

FIGURE 2-2 Teardown sequence

The exact diameter of a particular gauge isn't important, but it's critical that you develop an intuitive grasp for what's an adequate gauge for specific applications. For example, I wouldn't consider using an extension cord made with less than 14AWG copper wire. For standard hookup connections in an amplifier or other relatively high current circuit, I use 22AWG wire. When I'm dealing with robotic sensors and other weight-critical connections, I use 28 or 32AWG, depending on the physical stability of the circuit.

Step 1

With the unit face down, verify that the battery holder is empty and remove the four screws securing the back cover.

Step 2

Tilt the LED bank so that it is perpendicular to the body of the main unit. Pull straight up to separate the back from the top half of the body. Put the empty top half of the unit to the side. Position what was the bottom of the unit face up, as shown in Figure 2-2a. You can see the white Fresnel lens dome, surrounded by five electrolytic capacitors, on the main circuit board.

Also note the globs of glue on the bottom edge of the circuit board, where the wires from the LED bank attach to the board. Because the bank can be tilted, there's added stress on the wires where they attach to the board. The glue helps distribute this stress, thereby minimizing the odds the wires will break.

Step 3

Remove the two screws securing the main circuit board. Pop out the LED bank from the chassis. Flip the main circuit board over, as shown in Figure 2-2b. With the 16-pin IC to the right, you can see two, three-terminal SMT devices, labeled *Q1* and *U2*, and perhaps twenty SMT capacitors and resistors.

I took this photo after spending 15 minutes removing gobs and strands of hot melt glue. The glue not only obscured the components but also huge blobs of solder around some of the components. This is an example of poor SMT work that may have been the result of poor manual soldering technique, a dirty board, or the use of an automated solder machine in which the temperature of the solder was inappropriate.

Step 4

Working on the SMT component side of the board, use an ink pen or miniature screwdriver blade to push out the four plastic tabs securing the white plastic Fresnel lens dome. Turn the board so that the dome faces up. Gently remove the plastic lens from the top of the board. In the newly exposed area, you can see the PIR and CDS (cadmium sulfide) sensors, as shown in Figure 2-2c. Note the reflective window of the PIR sensor and the serpentine tracing on the face of the CDS sensor.

Step 5

Expose the second circuit board that forms the LED bank. Remove the black elastic bands from the stubs on either side of the bank. Pry open the case to reveal the circuit board holding four LEDs, four leaded 0.25W resistors, and the silver reflective assembly, shown in Figure 2-2d.

Layout

As revealed in the teardown, the layout consists of a 4.5VDC power source, a main board with sensors and other active components, and a small board with LEDs and resistors.

Now is a good time to hone your circuit tracing and component identification skills. Start by tracing the power and sensor signal paths with your ohmmeter. Next, insert three AAA cells in the battery pack and put the device through its paces.

What are the current and voltage demands when the device is on standby? What changes when the device is triggered? Any idea of what the three active components on the SMT side of the board are for? Try your best before continuing with this text.

Components

I have to admit that this teardown almost turned into a black box investigation. I spent several hours on the Web trying to find the datasheet on the 16-pin TL0001S integrated circuit. Initially, the best I could do was a spec sheet written in Chinese from the manufacturer in Taiwan.

In a parallel effort, I spent several hours trying to reverse-engineer the chip, working backward from the sensors and external resistors and capacitors. Then, by chance, I found a cross-reference to the BIS0001 IC, and then I was home free with a datasheet written in English. This is a good illustration of the value of looking for cross-references and equivalent components when you're hunting for a spec sheet.

TL0001S (BIS0001) IC

The TL0001S, which is equivalent to the BIS0001, is a special-purpose chip designed expressly to interface with PIR sensors as part of a motion-sensing circuit. In short, this is a plug-and-play chip designed for one thing. I checked out two other makes of PIR motion-sensing devices in my home, and both use the BIS0001 chip. The advantage of this chip over general-purpose chips and discrete components is vastly reduced component count and ease of manufacturing. Instead of mounting two or three op amp chips, a logic chip, three interface transistors for the sensors and output, and a host of bypass and biasing resistors and capacitors, a manufacturer can offer the same functionality with a single, affordable, space-saving chip.

The 16-pin TL0001S is packaged as a small outline integrated circuit (SOIC), as shown in Figure 2-3. Like the DIP (dual inline package) described in Chapter 1, pin 1 is marked with an indentation or other mark on the package top, and the pin numbering is counterclockwise from the top of the chip. This SOIC is about half the size of a 16-pin DIP.

The TL0001S contains five CMOS operational amplifiers, control logic, and dedicated interfaces for a PIR sensor, CDS sensor, and output relay or transistor. You can see in Figure 2-3 that the external component count is minimal. Most of the external components deal with biasing and control. You can modify the sensitivity, time on after motion ceases, and other chip parameters by adjusting resistor and capacitor values.

Two important parameters to consider are that the chip operates from 3 to 5VDC and can sink only up to 10mA. This is more than enough current to drive a single transistor or a solid-state relay, but it's not a typical mechanical relay and certainly not a bank of LEDs. If you look up the spec sheet for the BIS0001 on the Web, you'll find the suggested circuit is surprisingly similar to the circuit used in our motion detector.

FIGURE 2-3 TL0001S/BIS0001 IC

Passive Infrared Sensor

The PIR sensor used in our motion-sensing light is another example of a special-purpose device that holds a formidable position in the market. The DS203 PIR sensor, by PIR Sensor Co., LTD, is one of the most popular PIR motion-detector sensors available. According to the datasheet, available from Futurlec (www.futurlec.com), behind the single reflective window (see Figure 2-2c) are dual sensor elements that are used to detect variations in radiation in the 5 to 14 micrometer (μm) range. External supporting circuitry for the three-terminal device is minimal. The device requires ground and power and presents its analog output on a single pin.

The PIR sensor has a number of impressive features, such as the use of dual-sensor elements to suppress interference from ambient temperature fluctuations. The sensor isn't perfect, however, especially when used in a high ambient humidity and temperature environment. High humidity is a problem because moisture in the air absorbs IR radiation from the source. Because the sensor responds to fluctuations in temperature, as the temperature of the environment approaches external body temperature, the IR signature from humans blends in with background radiation. In both cases, the range of the sensor suffers.

Fresnel Lens

The dome-shaped Fresnel lens is semi-opaque to visible light, which reduces false triggering from ambient visible light sources, but it is transparent to IR radiation. The annular patterns on the dome's inner surface, visible in Figure 2-4, are

FIGURE 2-4 Fresnel lens close-up showing annular patterns

responsible for focusing the IR light on the window of a PIR sensor mounted at the focal point at the base of the dome.

If you've worked with a plastic Fresnel pocket magnifier, you know the main characteristics of a Fresnel lens compared with a conventional spherical lens. A Fresnel lens is generally thinner and lighter, made of inexpensive plastic, and yet provides significant magnification. The downside of a Fresnel lens is that the boundary between annular sections obscures the magnified image. That's why you don't see many camera lenses or reading glasses based on a Fresnel lens. However, in our application, image quality is irrelevant. We simply want to focus IR radiation within a specific area onto the window of a PIR sensor, and the inexpensive Fresnel lens does this superbly.

CDS Sensor

The CDS sensor, sometimes referred to as a photoresistor, is probably the most common ambient light sensor in use today (see Figures 2-2c and 2-5). The resistance of the CDS sensor in our motion detector changes from more than 1 million ohms (MΩ) in relative darkness to less than 100 ohms (Ω) in full sunlight. From a component perspective, you can think of the sensor as a two-terminal variable resistor. CDS sensors have been supplanted in many applications by more sensitive and responsive phototransistors.

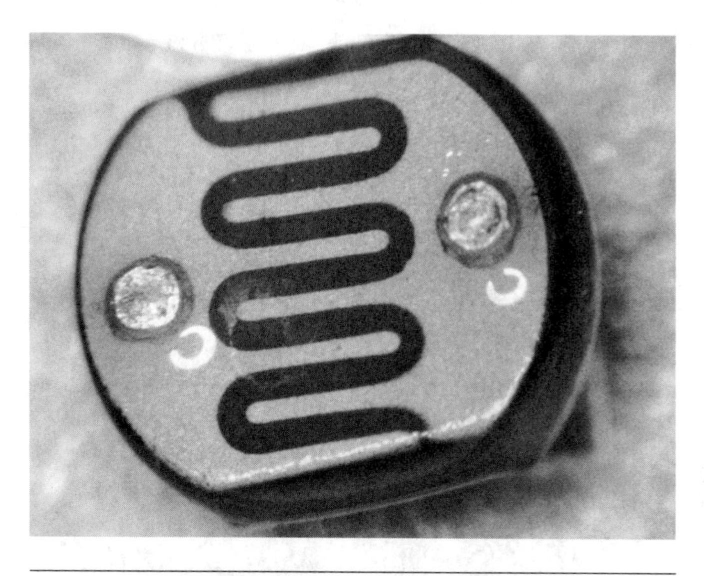

FIGURE 2-5 CDS sensor

7130 Voltage Regulator

The three-terminal 7130 is a 3V at 30mA voltage regulator. It can handle input voltages more than three times that supplied by the three AAA batteries used in our motion detector. More important, this is a *low-dropout regulator*, meaning that the regulator can function with just over 3V input from a nearly depleted battery.

As shown in Figure 2-6, the regulator is packaged in a 3-pin SOT-89 case, with input voltage on the middle pin, which is connected to the tab. In keeping with the SOT-89 standard, the top pin in the figure is connected to ground and the 3.0V output is taken from the bottom pin. If you trace the power from the battery, you'll find one wire to the voltage regulator input and another that goes directly to the bank of LEDs.

NPN Transistor

Near the edge of the main board, between the 16-pin chip and the wires leading to the small circuit board, is Q1, an NPN transistor in an SOT-323 package (see Figure 2-7). The SMT transistor is used to switch on power to the bank of LEDs, as a function of the output of the BIS0001 chip.

As with the voltage regulator, you can use the standard package configuration to identify the transistor leads. With the transistor oriented with the single, center lead to the left, as shown in Figure 2-7, the base lead is on the bottom right, the emitter on the top right, and collector center left. The SOT-323 package is a miniature version of the SOT-23 package, which you'll see plenty of in subsequent teardowns.

If you study Figure 2-7, you'll appreciate my regard for the quality of soldering in this device. The solder pad for the emitter lead of the transistor is nearly touching the trace immediately above it, thanks to the surplus solder on the pad. The

FIGURE 2-6 7130A-1 voltage regulator

FIGURE 2-7 Q1 SOT-323 package

numerous speckles on the board are remains of large globs of glue that I removed with forceps.

LEDs

White LEDs, once the rage in a world limited to greens and reds, are now available for pennies. The four LEDs used in this motion detector are moderate intensity LEDs of the type typically found in portable reading lights. Each of the four LEDs and its 27Ω series resistor is connected in parallel.

LEDs emit light when forward biased around 2V, with the exact voltage dependent on the diode design. Except for high wattage LEDs, connecting an LED directly to a battery typically results in a very brief flash of light and a burned-out LED. In our motion detector, the 27Ω series resistors limit the current through each LED to about 25mA, which is typical for white-light LEDs used for illumination.

You can calculate the value of the current-limiting resistor for any LED and voltage source with the following equation:

$$R = (VS - VD)/I$$

where *VS* is the source voltage, *VD* is the diode forward voltage drop or operating voltage, and *I* is the operating current. You'll also find several good LED current limiting resistor value calculators on the Web.

Capacitors

The motion detector circuit relies on both leaded electrolytic and SMT ceramic capacitors. The relatively large, leaded electrolytic capacitors are probably used in this application because they're less expensive than their SMT counterparts.

If you want to practice reworking an SMT board, the ceramic capacitors are a great target for removal and replacement.

Resistors

The resistors on the main board are standard tolerance, 1/8W, SMT resistors. The 0.25W leaded resistors are used as current-limiting resistors for the LED bank instead of more space-efficient SMT resistors because of the higher wattage requirement.

Battery

The motion detector draws about 1mA when idle and about 138mA with the LEDs energized. Given a typical alkaline AAA is rated at about 1 ampere-hour (Ah), how long will the battery last if the unit is perpetually idle? What about perpetually on? Here's the basic formula for amp-hours:

$$Ah = I \times T$$

where *I* is the current in amps and *T* is the time in hours. For example, a 1Ah battery can theoretically supply 1A for 1 hour or 500mA for 2 hours. In practice, the formula and rating break down at high current and long time intervals. It's safe to say that a 1Ah AAA cell can supply 100mA for 10 hours, but not 100A for 0.01 hour. Similarly, a cell has a finite shelf live. A 1Ah AAA cell won't supply a 1μA circuit for a century.

If you want to verify your Ah calculations, place your ammeter probes across the collector and emitter of Q1. Use a timer and note the starting time and the time when the LEDs are no longer illuminated. If you want to shorten the time of the experiment, use generic carbon zinc cells with a rating of 0.4Ah. On the Web, you can find the rating of standard cells by chemistry, voltage, and manufacturer on Digi-Key (www.digikey.com) or other electronics parts dealers' sites.

How It Works

The sensor technologies that form the basis for most motion-activated lights, cameras, and security systems include active IR, active ultrasound sensing and ranging, and passive IR (PIR) detection. The operation of active IR and active ultrasound sensors is similar to the operation of radar in that a beam of energy is transmitted and the reflection is received and analyzed for changes in timing, amplitude, or phase. If the time between successive reflections decreases, this suggests that the distance between the sensor and an object in the sensor's active area is decreasing. If the amplitude or phase of the reflected signal changes, this suggests that something or someone is interrupting the transmitted or reflected beam.

PIR detection is based on relative changes in the temperature of an area being monitored, based on fluctuations in the level of infrared radiation. Everyone and everything not at absolute zero (–459.67°F or –273.15°C) emits radiation, and some of this radiation is within the infrared spectrum, which spans the range of about 750 nanometers (nm) to 100μm. Different objects and biological systems have different spectral fingerprints, just as objects vary in visible light absorption or perceived color. At normal body temperature, our IR fingerprint has a peak around 10μm. As such, an IR detector that is sensitive to 10μm radiation should be good for detecting the presence and movement of humans.

Each of these methods of motion detection has practical benefits and drawbacks. Passive detection is characterized by relatively low power consumption, for example. A battery-powered PIR-based motion detector can operate for months between battery changes. Moreover, a battery-powered unit is more desirable than an AC-powered monitor in a security system, because AC power can be intentionally disrupted.

Active approaches to motion detection generally have longer range and greater sensitivity, and they are more difficult to evade. For example, because a PIR detector is sensitive to rapid changes in relative temperature, a burglar could avoid detection by moving slowly through an area wearing a heavily insulated jacket and hood. Active IR motion detectors, in contrast, aren't fooled by heat-absorbing insulation. A new object in the path of an active IR pulse results in a new reflection, which can be used to trigger an alarm.

Active ultrasound methods are also impervious to ambient light, and so can be used outside in the daylight. In addition, the range of active ultrasound motion sensors is 30 feet or more—more than twice the range of a typical IR motion detector.

Figure 2-8 shows the simplified schematic of our motion-activated light. Note the package outlines for the 7130 fixed voltage regulator (SOT-89) and the much smaller Q1 (SOT-323). The package outlines are not drawn to scale.

Starting at the upper left of the schematic, the battery supplies 4.5VDC directly to the 7130 3.0VDC fixed voltage regulator and the physically separate bank of LEDs. No on-off switch or fuse lies between the battery and other components.

One reason for the lack of fusing is the relatively low power available from the three AAA cells. Another is that the 7130 voltage regulator is virtually immune to damage from component failure or battery voltage reversal. The fixed voltage regulator can tolerate a short on the output and will protect the components from reverse voltage. A 100µf at 10VDC electrolytic capacitor is used on the output of the voltage regulator to enhance regulation.

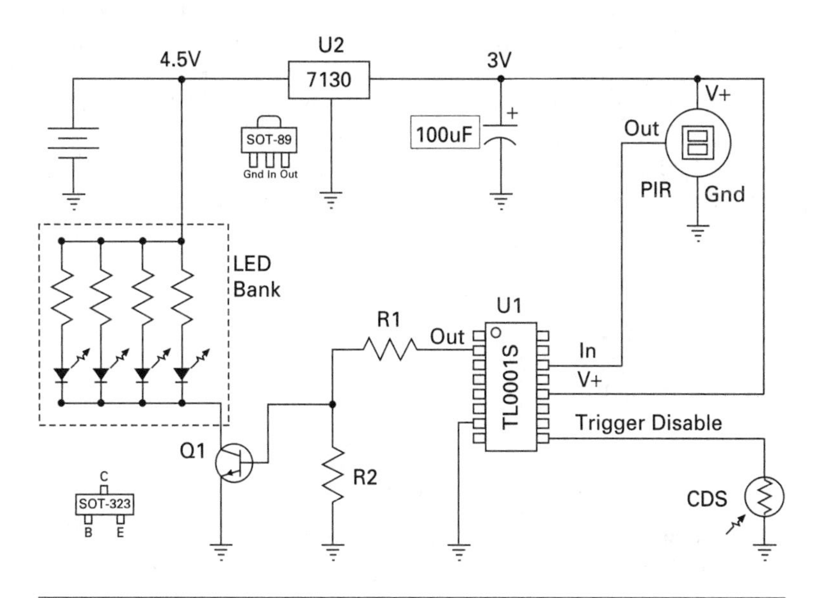

FIGURE 2-8 Simplified schematic of motion-activated light

The circuit lacks a manual on-off switch because the manufacturer can get away without installing one, thereby saving the engineering, manufacturing, and component cost. There's also one less item to fail. A switch isn't absolutely necessary because of the low standby current requirements of the CMOS TL0001S chip and PIR detector, which amounts to only a few microamps. Even when both are triggered, the two components draw less than 100μA.

The other components that could draw down the battery, the bank of LEDs, are similarly insignificant sources of current drain. The switching resistor (Q1) is normally reverse biased, with the base at ground potential, resulting in an essentially open collector-emitter connection.

Although there is no manual on-off switch, the CDS sensor performs an equivalent, automatic function. When the ambient light level is high, the resistance of the CDS is low, and the normally high trigger disable pin of the TL0001S chip is pulled low. This shuts down the chip, making it unresponsive to input from the PIR sensor.

When it's dark, the resistance of the CDS is high, which allows the trigger disable pin of the TL0001S to return to its normally high state. The chip is now responsive to signals from the PIR sensor. When someone walks within the detection range of the PIR sensor, the person causes a fluctuation in the output of the PIR sensor on the order of a few millivolts (mV). This signal is amplified and processed by the logic within the TL0001S chip.

If the fluctuations in PIR output are significant, the normally low output pin of the TL0001S is driven high, and current flows through R1 and R2. The positive voltage drop across R2 forward biases Q1, providing a low resistance ground connection for the four LEDs, which draw a total of 40mA at 4.5VDC.

When the fluctuating heat source moves out of the detection range of the PIR sensor, the sensor output signal will become quiescent in 20 seconds. The TL0001S output pin goes low, as does the base of Q1. Without forward bias, the collector-emitter path returns to a high resistance state, and no current flows through the LEDs. The resistor (R2) from the base of Q1 to ground serves to guarantee the base voltage doesn't float above ground due to noise.

Mods

Because the dedicated chip simplifies the overall circuit, mods are limited but also greatly simplified.

Narrowed Trigger Zone

The quickest mod is narrowing the trigger zone. Let's say you have a hallway with exits to the left and right, and you want to limit triggering to when someone walks to the right. Assuming the 40-degree cone of detection is too broad, you can easily narrow the detection angle to a few degrees by fitting a tube over the Fresnel dome. Take the cardboard tube from a roll of aluminum foil or plastic wrap, cut a 2-inch length with a blade, and paint the inside flat black to absorb stray IR and visible light. Paint the outside silver or the color of your particular motion detector. Mount the tube around the dome with tape or a dab of silicone glue.

You might want to try a few different tube lengths before you glue the tube in place. The longer the tube, the narrower the detection zone. I've tried tubes up to 4 inches long with a successful detection range of about 5 feet.

External Device Controller

You can use the motion detector to trigger anything from a 120VAC light to a speech synthesizer on a mobile robot. Simply take the two wires leading to the LED bank and run them to a solid state or electromechanical relay capable of handling the voltage and current in your circuit. With an isolated output, you can connect anything you like to the output contacts of the relay, without regard for the components and circuitry in the motion detector.

If you opt for an electromechanical relay, make certain the motion detector can meet current demands of the relay coil. Keep the load on Q1 at or below 160mA, which is the load imposed by the LED bank.

You could simply connect the wires intended for the LED bank to the load of the second device. Assuming the device runs off of DC, you can use a battery supply up to about 24VDC, the maximum input to the 7130 voltage regulator. You may have to replace R1 and R2 with a 25KΩ potentiometer, however, because the bias setting will be different with a higher collector-emitter voltage.

Adjustable Light Sensitivity

You can adjust the ambient light sensitivity, time delay, and other basic parameters of the TL0001S by changing the values of the external resistors and capacitors. See the datasheet documentation for details. However, if you don't want to work with SMT components, you can make simple mods within minutes. Want the detector to work in a higher ambient light condition—maybe even daylight? Simply remove the dome and tape over the CDS sensor with black electrical tape. If you want a permanent fix, cut one of the leads to the CDS sensor and replace the dome.

Chapter 3

Digital Bathroom Scale

Digital bathroom scales, such as the one shown in Figure 3-1, have evolved significantly since they were introduced in the consumer market more than a decade ago. My first all-electronic digital scale had an aluminum chassis about 4 inches thick, weighed 10 pounds, and had a 12VDC external power supply. The cost for this space-age marvel was more than $200. Unfortunately for the average consumer, the market was flooded with cheap knockoff scales that had a digital readout but retained the relatively inaccurate spring-based weighing mechanism.

Today, inexpensive, svelte, feature-laden digital scales with features from body composition to Bluetooth connectivity are the norm. It's actually difficult to find an old-fashioned analog spring steel bathroom scale. After all, why would anyone want a scale with poor repeatability that's based on a mechanism that's destined to rust in a constantly humid environment? Admittedly, balance beam scales—like those used

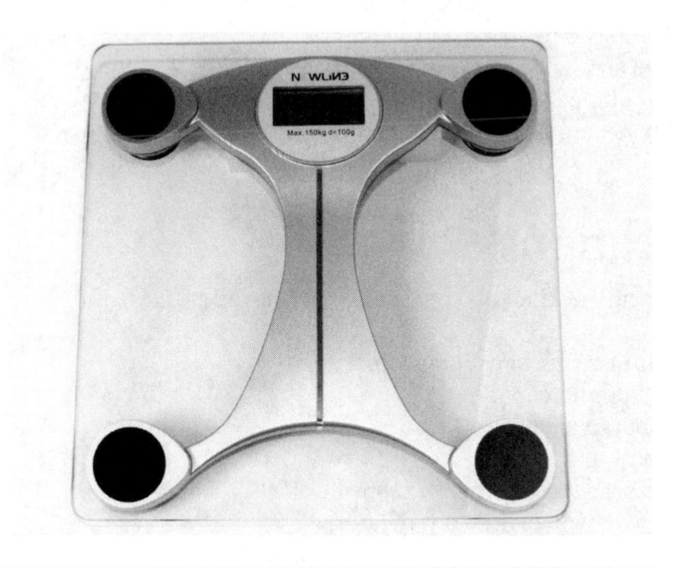

FIGURE 3-1 Newline Well Balance bathroom scale

in doctors' offices—are accurate and can last indefinitely, but they take up valuable space in a home bathroom. It's easy to slide a modern digital scale under a cabinet when it's not in use.

In this chapter, I'll tear down a Newline Well Balance Digital Bathroom Glass Scale (www.newlinescale.com). This scale doesn't talk to you in synthesized speech or indicate your percentage of body fat, but it does measure your weight in pounds or kilograms quickly and with good repeatability. In addition to exploring the overall design of this Chinese-manufactured, $25 scale, we'll review the fundamentals of strain gauge sensors and investigate the scale's unique hibernation and wakeup system.

Highlights

The star of this teardown is the strain gauge, which is the basis for accurate, reproducible measurements. The scale also employs a unique vibration sensing mechanism to activate the scale.

This teardown also features a mystery microcontroller entombed under an unmarked black blob of epoxy. This form of encapsulation is a common method of protecting naked ICs that are wired directly to the circuit board. In effect, the underlying circuit board is part of the IC. Epoxy encapsulation is an economical mounting option for manufacturers that work with multiple custom chips but a pain for curious hobbyists and a speed bump for competitors attempting to reverse-engineer a chip.

During the teardown, note the following:

- The black blob entombing a mystery microcontroller
- The piezoelectric transducer
- The soft, spongy connector between the LCD (liquid crystal display) and the circuit board
- The location, structure, and wiring of the strain gauges

Specifications

Following are the key specifications offered by Newline:

- Strain gauge sensor system
- 330 pound capacity
- Foot tap switch
- 1 inch LCD
- Auto zero resetting/auto power off
- Low power/overload indicator
- Powered by a pair of 3V CR2032 lithium cells

Of note is that there is no mention of accuracy on the Newline web site or in the material accompanying the scale.

Operation

If you check the user feedback on Amazon.com for this scale, you'll see that many people have trouble operating it. This seems paradoxical at first, because this scale is about as simple as it gets. The only setup is selecting pounds or kilograms, using the switch on the underside of the scale. Unlike more advanced scales, there's no age or height data to enter, no memory selector to store the weight for different family members, or any of the other popular features. However, there's no manual on–off switch—the source of the complaints.

According to the manual, to activate the scale, you simply tap it with your toe. In practice, you have to lift the edge of the scale an inch above an uncarpeted floor and then release the scale. If you were sufficiently rough in handling the scale, the LCD display comes to life, indicating that you can step on the scale. I haven't had any problem with the scale, but subjecting the scale to a 1-inch drop onto a ceramic tile floor every day can't be good for the strain gauges.

Teardown

You should be able to complete the teardown of the scale, illustrated in Figure 3-2, in about 15 minutes. The work entails extracting the load cells—the strain gauges and their mechanical mounts—from the four feet and the display and removing the processing unit from the head of the scale.

Tools and Instruments

If you're following along at home with a digital scale, you'll need a good magnifier and forceps in addition to the usual screwdriver, soldering iron, and multimeter. A digital multimeter (DMM) with a high-resolution, accurate ohmmeter function will help when we examine the strain gauges in detail.

Step by Step

Before you dismantle the scale, it's a good idea to test the scale and study the front display, shown in Figure 3-2a, so that you can appreciate the underlying technology. How's the repeatability of your scale? If you stand to one corner of the scale, is the reading different from when you stand with your weight evenly distributed among the four feet? If you have the same model of Newline scale, can you easily activate it?

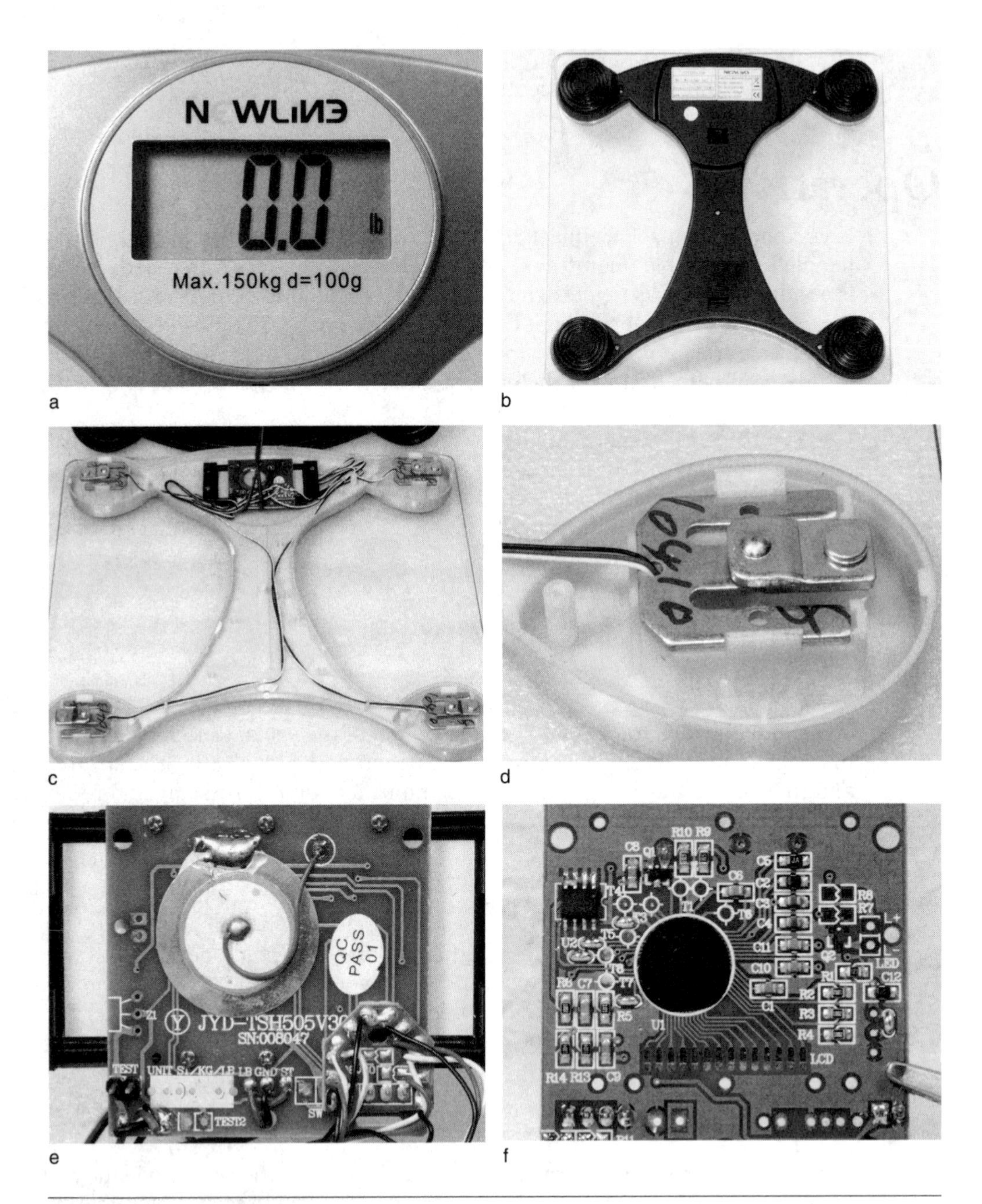

FIGURE 3-2 Teardown sequence

Step 1

Remove the two 3V lithium cells. Place the scale face down, as shown in Figure 3-2b, and slide open the battery compartment in the lower half of the scale. Remove the two CR2032 cells and set them aside; you'll need them later.

Step 2

Remove the back of the scale. Remove the five Phillips-head screws from the back—one in the center and one over each of the four arms. Gently pry the black plastic back from the white plastic front, working from the end farthest from the display. Hinge the back away from the remainder of the scale, as shown in Figure 3-2c, taking care not to rip the cables to the pounds/kilograms switch and the battery pack that remain with the black plastic back. Note that the battery pack is wired so that the lithium cells are in parallel, and that the voltage supplied to the processing and display assembly is 3V.

Step 3

Extract the load cells from each of the four feet. Each load cell, such as the cell shown in Figure 3-2d, snaps out of place with a little finger pressure. Glued to the bottom of each load cell is a strain gauge that's connected to the main control unit via a red, white, and black wire.

Step 4

Examine and test each load cell. The load cells in this scale are three-pronged steel devices. A thin foil strain gauge proper is glued to the underside of each cell, and the wires are held in place with a dab of silicone sealant.

Insert the lithium cells in the scale back and tap on the display to activate the scale. Apply pressure to one of the load cells by pushing down on a load cell with the palm of your hand. A weight value should appear on the display.

Now test the load cells. Can you tell if the weight measurement is additive? You have to twist or add pressure to a load within a few seconds of activating the scale or the scale will power down automatically.

Repeat the process with an ammeter in series with the 3V battery pack to determine the current load. You'll have to cut or unsolder one of the wires leading from the battery pack and insert your ammeter in series with the wire. I read 4mA when the display is active and about 5μA when the scale is in standby mode. Remove the lithium cells from the back of the scale when you've finished experimenting.

Step 5

Remove the LCD panel assembly by lifting it straight out of the scale front. Note the main circuit board attached to the display back, as shown in Figure 3-2e. Note the prominent, quarter-sized piezoelectric transducer soldered to the back of the circuit board. The small, six-sided connector board in the lower-right corner of the figure is connected to each of the four load cells. Remove the four-pin connector now.

The wires from the battery pack enter at the lower-left corner of the board. Along the lower edge of the board, between the power and strain gauge connector, is the pounds/kilograms switch.

Step 6

Remove the LCD from the circuit board. Remove the six screws holding the board in place. If you didn't remove the four-pin strain gauge connector in the lower-right corner of the board, you won't be able to get at one of the screws. The LCD panel should fall away from the board. Carefully store the LCD panel and the soft pink connector for use later.

Step 7

Examine the back of the board. Flip the board over and orient it to match Figure 3-2f. To the left of center is a black blob—a 1/2 inch–diameter mound of epoxy that protects a microcontroller and associated wiring. You'll see an SMT (surface mount technology) transistor and 8-pin EEPROM (electrically erasable programmable read-only memory) chip in the upper-left corner of the board. About a dozen SMT resistors and capacitors are distributed to the left and right of the blob. In the lower center of the board are the multiple fingers that mate with the LCD module through the soft connector.

Layout

The teardown revealed a simple and straightforward layout. The four peripheral strain gauge sensors, the weight system switch, and the battery pack are all connected to the central processing and display unit. The double-sided board has numerous markings for test points and markings for components used to drive an LED, which is apparently for some other model of scale.

Components

The real stars of this teardown are the four strain gauges that make this all-electronic scale possible. The flexible coupler between the LCD and the circuit board is worth exploring as well. Although the onboard microcontroller is unmarked, we can deduce the functionality from the product specifications and other hardware on board.

You can use a few SMT resistors and capacitors on the board to practice removing and installing SMT components. The resistors are stamped with values—see Appendix A if you need to brush up on your resistor identification. Without an official schematic, the quickest way to determine capacitor values is to use a capacitance meter with a forceps probe.

Microcontroller

The blob of black epoxy is a microcontroller, because it has analog-to-digital conversion capabilities, on onboard oscillator, timers, and a driver for the LCD. The next best alternative, a microprocessor, doesn't fit the specifications. Microprocessors may be computationally superior to microcontrollers, but they have limited input/output capabilities.

Given the functionality of the scale, we can assume the microcontroller provides the analog-to-digital (A–D) conversion of the strain gauge signals, performs the pounds-kilogram units translation, shuts down the scale after a period of inactivity, powers up the scale when the scale is jostled, and drives the LCD. The peripheral EEPROM, discussed next, suggests that this is an 8-bit microcontroller.

If you want to see what's under the blob, you can use two practical ways to get at the underlying chip: either chemically etch the epoxy or soften the epoxy with an SMT hot air pencil and remove the crumbly mass with a small blade. A third approach that seems to be popular on the Web is to use a rotary tool and rotary file to wear away at the blob until you hit something interesting. In my experience, getting physical with the epoxy results in a small pile of unrecognizable pieces of silicon and epoxy.

If you want to maintain the integrity of the small gold or copper wires between the chip and the circuit board, you can try to dissolve the blob with ATTACK epoxy solvent. (An 8-ounce container is available through online retailers such as Amazon.com for about $10.) The solvent works exactly as advertised, but it's also extremely hazardous. Don't even think of using it indoors, around life forms you value, or without skin protection. The U.S. Postal Service won't even ship it. And I doubt that the board will be useable after your apply the solvent—but it will look good.

The approach I recommend for novices is to simply heat the blob to about 900°F with a hot air pencil. At that temperature, the epoxy behaves like soft clay, and you can easily pick away at it with a hobby knife, scalpel, or other small, sharp instrument. Some of the traces may be removed during the process, but the chip should come out unscathed.

I used the thermal approach to create Figure 3-3, which shows the microcontroller chip sitting on the much larger rectangular pad. To improve the clarity of the image, I superimposed a high-resolution photo of the newly uncovered area onto the original circuit board image shown in Figure 3-2f. Otherwise, the images are untouched.

Visualizing the silicon chip probably doesn't tell you much about the architecture or functionality of the microcontroller. However, a microcontroller designer can tell a lot from viewing the chip, because each section–the mathematical unit, onboard RAM (Random Access Memory), and various registers–has a characteristic appearance. For our purposes, getting a glimpse of the microcontroller is simply part of the fun of a teardown.

FIGURE 3-3 Circuit board after removal of epoxy blob

EEPROM

The ATMEL 718 24C02BN EEPROM provides 2K of memory, organized with 32 pages of 8 bytes each. This is an 8-bit component, strongly suggesting that the microcontroller is an 8-bit device as well. The microcontroller apparently requires external memory to store the program code required to direct the operation of the scale. There is no evidence that the EEPROM can be erased and reprogrammed by the microcontroller in the field.

The specifications of the chip include a two-wire serial interface and 2.7VDC operating voltage. The two-wire interface agrees with my findings on the board, in that only a few wires connect the microcontroller and EEPROM. The operating voltage is compatible with the supply voltage available in the scale.

It's always a good idea to check the specs against the real-world configuration of a device. In some instances, you may have mischaracterized a device or made incorrect assumptions about the circuit configuration.

Strain Gauges

The weight sensors on this scale are thin foil strain gauges. Each of these gauges is glued to a metal, weight-bearing structure in each of the scale's four feet. Recall that a strain gauge and its supporting, weight-bearing structure are referred to as a *load cell*. Review photos of the load cell in Figure 3-2d, and, if you're following along with your own teardown, examine each load cell carefully. Note how each strain gauge is positioned exactly over the pivot point on the metal support where a metal extension is riveted to the main body.

Both the orientation and placement of each strain gauge on its supporting structure are significant. You can get a better appreciation of the placement of each strain gauge if you remove the white silicone goop from the underside of each load cell with your fingernails. Figure 3-4 shows a strain gauge still glued to the bottom of a weight-bearing structure. The wires were removed for clarity. The three prongs of the load cell open to the right, and the wires normally connected to the sensor, from top to bottom, are white, red, and black.

The metal extension on the support has the effect of subjecting one side of the strain gauge to compression and the other to elongation. This deformation is critical, because strain gauges respond to strain. Recall from basic physics that *strain* is the fractional change in length of a body that is subject to an applied force. Each of the strain gauges in our scale is oriented with the zigzag pattern running from left to right, which is the direction of strain. When you stand on the scale, the strain gauge element on the outside membrane is stretched, the metallic foil pattern is thinned, and resistance increases. Simultaneously, the inside element is compressed, the foil is thickened, and resistance decreases. Because of the zigzag pattern, the relative change in resistance is multiplied relative to what a single conductor would exhibit.

You can confirm these assertions with your DMM. If you check the resistance between the three solder pads of a strain gauge, you'll find that each side has a resistance of about 1KΩ, or 2KΩ when both sides of a strain gauge are measured in series. If you return the load cells to their original housings and measure the resistance between the three leads, you'll find that with no load, the resistance from

FIGURE 3-4 Foil strain gauge glued to metal pod

the center to either side of a strain gauge is 1KΩ. When you stand on the scale, the resistance of the outer side of each strain gauge should increase by 2 to 4Ω, and the resistance of the inner side should decrease by 2 to 4Ω. Clearly, the position of each strain gauge in its metal pod is such that one side is subject to compression and the other elongation. This finding has major implications, as discussed a bit later.

The metal foil strain gauges used in this scale are popular because they're inexpensive and relatively robust. However, if this were a laboratory scale instead of a bathroom scale, the strain gauges would most likely be made of semiconductors, which are more sensitive, more fragile, and more expensive than metal foil strain gauges. Both types of strain gauges are affected by ambient temperature. However, metal foil strain gauges can be made relatively immune to temperature changes by using alloys that minimally contract or expand with such changes.

Piezoelectric Transducer

The piezoelectric transducer soldered to the back of the main circuit board is a generic component typically used with an oscillator to create a buzzer. In this application, the transducer is used as a sensor. Vibrations caused by dropping the scale are converted to electrical impulses that are used to activate the scale.

Can you think of why the designers decided to add another sensor to the scale when they already had four strain gauges to detect someone stepping on or dropping the scale? It's probably because monitoring the strain gauges requires significant processing power and energy from the battery. It's more efficient to have the microcontroller hibernate until awakened by a pulse from the piezoelectric transducer. An even more efficient, but less attractive, alternative is a mechanical switch, but this solution has limited market appeal.

J3Y Transistor

The output of the high impedance piezoelectric transducer is insufficient to drive the microcontroller directly. The J3Y transistor, an NPN transistor in an SOT-23 case, provides the needed amplification and impedance matching for the microcontroller. The transistor, which is equivalent to the S8050LT, has a collector current rating of 0.5A and a dissipation of 0.3W.

In case you're wondering, I don't have an encyclopedic knowledge of transistors. I found the current rating, dissipation, and other specifications of the J3Y transistor the old-fashioned way—by searching the Web. I've had the best luck with Hong Kong supply houses that list transistors and their equivalents. Often the equivalent is easier to locate on yet another supply house list. Make certain you include the full transistor designation on your search. For example, if you search for "8050 transistor," you may find the datasheet for the NPN transistor in a leaded SO-92 case, which has a dissipation of 1W.

Even without a datasheet, identifying the basic transistor parameters is relatively simple. For example, I know the identity of the leads from standard small outline transistor (SOT) packaging. Looking at the component with the single lead on top and two leads on the bottom; the collector is on top, the base lead is on the bottom left, and the emitter is on the bottom right. Following this pinout, the emitter is grounded, the base is connected to the piezoelectric transducer, and the collector is connected to a microcontroller line. The collector is also connected to the 3V battery supply through a resistor.

As discussed in Chapters 1 and 2, given a transistor in an SOT-23 case without a datasheet, I know the emitter, base, and collector leads. It's up to you to decide whether the device is an NPN or PNP transistor. One way to make this determination is to examine the transistor with a multimeter. If you have an NPN transistor, with the positive probe on the base lead, you should read relatively low resistance to the emitter and collector leads. If the resistance is several million ohms, it's a PNP transistor. I prefer an analog ohmmeter to a DMM for checking transistors.

The second method of determining transistor type is to examine the circuit. Given that the emitter is grounded, odds are we're dealing with an NPN transistor. Recall that an NPN transistor is forward biased when the emitter is negative and both base and collector are positive. The forward bias condition for a PNP transistor is the inverse—the emitter is positive and both base and collector are negative. In our circuit, the emitter is grounded and the collector is positive because of the resistor to the 3V supply—clearly conditions for an NPN transistor.

LCD

As we discovered during the teardown, no ribbon cable or soldered connector lies between the circuit board and LCD. Instead, a pink Zebra Elastomeric Connector is friction-fit between the fingers on the circuit board and the etched edge connectors on the LCD. Figure 3-5 shows the conductive fingers on the edge of the LCD panel juxtaposed with the edge of the Zebra connector. The bottom third of the figure shows the LCD panel.

Whoever named the Zebra connector must have had a powerful microscope. The stripes, formed by layer upon layer of insulator alternating with carbon, silver, or gold conductor, are barely visible in Figure 3-5. These layers run perpendicular to the long axis of the connector, so that each conducting layer connects one side of the connector to the other. Think mile-high tiramisu, with each cake layer a conductor and each creamy layer an insulator.

To give you an idea of scale, the spacing between conductive fingers of the LCD in the figure is about 1mm. As you can see, there are about ten stripes or conductor-insulator layers per finger, which makes each stripe on the Zebra connector about 0.1mm thick.

A major limitation of the elastomeric connection is that it provides a relatively high resistance path—on the order of 1KΩ—for low-current applications. This is

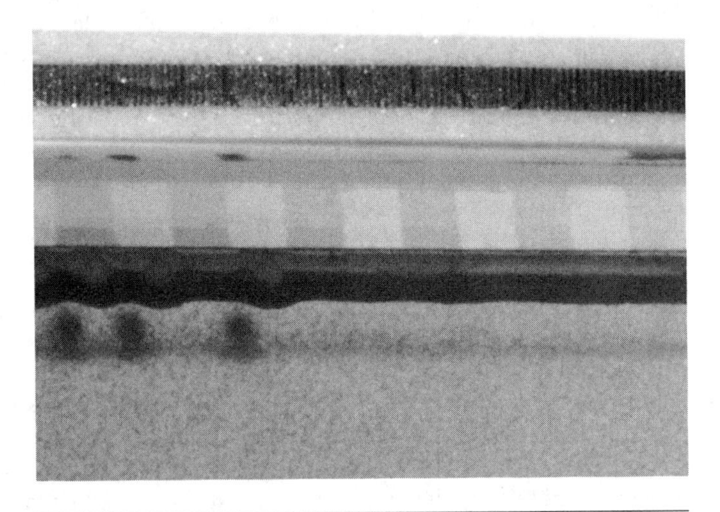

FIGURE 3-5 Zebra Elastomeric Connector (top) and LCD edge connectors (middle)

perfect for a low-current LCD, but it wouldn't work for, say, a high-current video board or electric motor connection. For more information on the Zebra connector, try the Fujipoly America web site, at www.fujipoly.com.

How It Works

A simplified schematic of the scale is shown in Figure 3-6. As previously discussed, two 3V lithium cells in parallel supply power. The microcontroller and EEPROM provide programmatic control for the automatic function and LCD display, and a Zebra Elastomeric Connector provides connectivity between the microcontroller and LCD. Note the pinout for the SOT-23 package, which applies to the transistor (Q1).

The trigger circuitry consists of the piezoelectric transducer (PT), NPN transistor (Q1), and pull-up resistor (RP). When the transistor is in the quiescent, nonconducting state, the positive supply voltage is supplied, via the RP, to the microcontroller input line. When the piezoelectric transducer is jolted, it produces an AC signal, the positive component of which forward-biases the transistor. When the transistor conducts, the input line to the microcontroller is pulled down to a few tenths of a volt. This change of state signals the microcontroller to emerge from standby mode and activate the LCD and bridge circuits.

Recall that an NPN transistor is forward-biased when the base and collector are positive relative to the emitter. Also note that, for clarity, the signal from

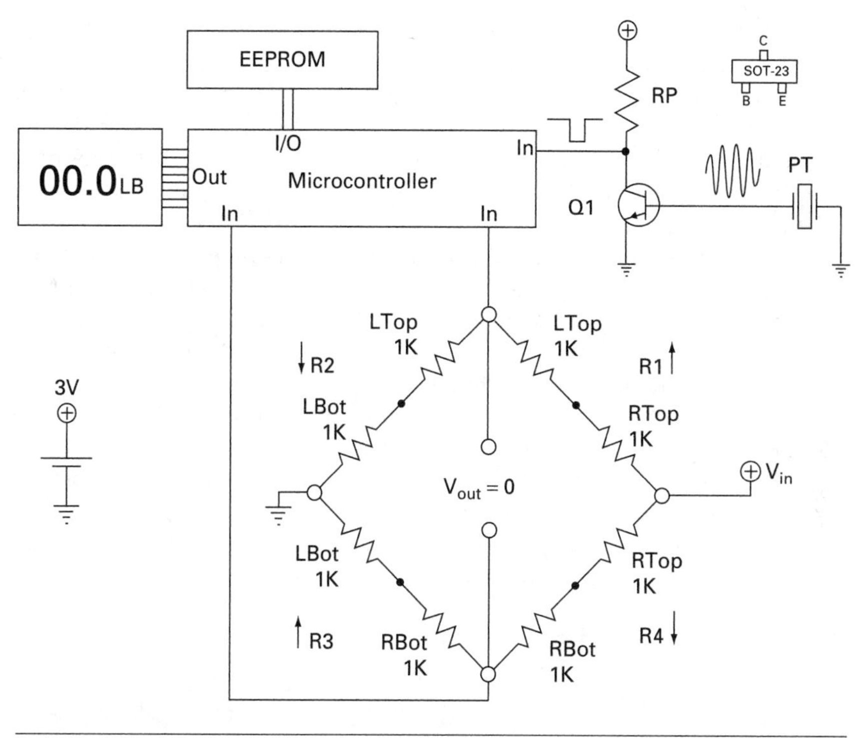

FIGURE 3-6 Simplified schematic of scale

the piezoelectric transducer and output of the trigger circuit are not to scale in the schematic. In addition, only one negative-going pulse is shown on the microcontroller line. In reality, there is a string of pulses, as a function of the severity of the jolt to the piezoelectric transducer. The microcontroller is presumably programmed to ignore activation pulses that closely follow the first pulse.

The most notable feature of the schematic is the resistance bridge formed by the four pairs of strain gauge segments in the load cells. This configuration of resistances, referred to as a *Wheatstone bridge*, produces an output voltage (V_{out}) that is sensitive to minute changes in the relative resistances of each leg of the bridge. This output voltage is sent to the microcontroller, which determines and displays the weight of the person standing on the scale.

The Wheatstone bridge consists of eight separate resistances, the contributions of the two sections of each strain gauge. At rest—that is, without load—the right-top (RTop), right-bottom (RBot), left-top (LTop), and left-bottom (LBot) strain gauges each contribute two 1KΩ resistances to the bridge. The names of each strain gauge correspond to the positions of the load cells in the scale. For example, the RTop strain gauge is located in the right-top foot of the scale, looking at the scale from above, as shown in Figure 3-1.

I labeled the sides of the bridge R1–R4, as shown in Figure 3-6, so that the RTop strain gauge contributed to R1 and R4. The side of the strain gauge that decreases in resistance with weight applied to the load cell contributes to R4, signified by the downward arrow adjacent to R4. The side of the strain gauge that increases in resistance with weight contributes to R1. The remaining three legs of the bridge are similarly configured so that each leg either increases or decreases with weight on the contributing strain gauges.

Now focus on the four sides of the bridge, R1–R4, each 2KΩ in value, with two parallel paths formed by R1–R2 and R3–R4. Hopefully, it's intuitive that the output (V_{out}) from the bridge is 0V, regardless of the input voltage (V_{in}). The same relative voltage drop appears across the voltage dividers R1–R2 and R3–R4.

To illustrate, if V_{in} = 10V, then the voltage drop across R1 = 5V and the voltage drop across R4 = 5V. Using ground as a reference, the junction between R1 and R2—the top corner of the bridge—is at 5V. Similarly, the junction between R3 and R4—the bottom corner of the bridge—is at 5V, relative to ground. Because the top and bottom corners of the bridge are at 5V relative to ground, there is no potential difference between the two corners of the bridge. That is, $V_{out} = 0$.

We can prove this with Ohm's law:

$$V_{out} = V_{in} \times [(R3/(R3 + R4)) - (R2/(R1 + R2))]$$

Assuming all resistances equal 1KΩ, the output voltage is

$$V_{out} = V_{in} \times [(1/2) - (1/2)]$$

$$V_{out} = V_{in} \times 0$$

Regardless of the input voltage (V_{in}), when the resistances in each of the four legs of the bridge are equal, output is 0V. As shown in Figure 3-6, the appropriate LCD display with a 0V input to the microcontroller is 00.0 pound. Of course, R1–R4 aren't exactly 1KΩ, and there will be some output from the bridge with no load. However, the microcontroller uses the initial output of the bridge at startup as 00.0 pound.

Now consider the Wheatstone bridge with a weight of about 173 pounds on the scale—my weight—as shown in Figure 3-7. As you can see, the bridge is unbalanced. Using the previous formula for output

$$V_{out} = V_{in} \times [(R3/(R3 + R4)) - (R2/(R1 + R2))]$$

$$V_{out} = V_{in} \times [(2006/(2006 + 1994)) - (1994/(2006 + 1994))]$$

$$V_{out} = V_{in} \times [(2006/(4000)) - (1994/(4000))]$$

$$V_{out} = V_{in} \times 0.003$$

Assuming an input voltage (V_{in}) of 10V, the output voltage (V_{out}) is 30mV, or about 10mV with an input voltage of 3V. While 10mV may not seem like a significant signal

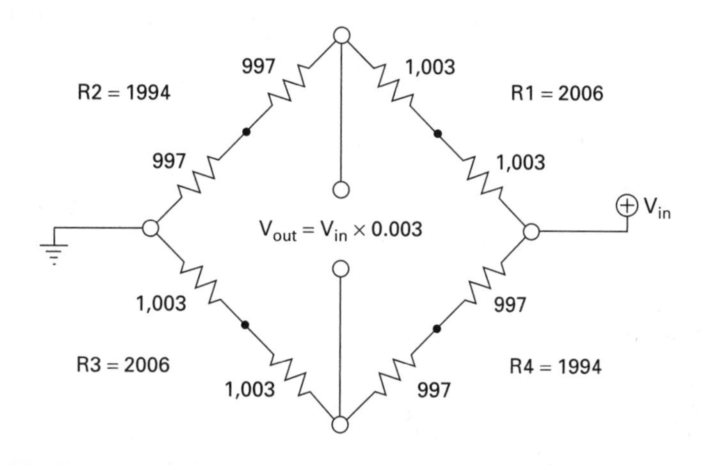

FIGURE 3-7 Wheatstone bridge with 173.4 pounds on the scale

level, it's sufficient for the microcontroller. Strain gauges with greater sensitivity are available for scales with a smaller weight range, such as a kitchen scale with a maximum capacity of a pound or two. In general, strain gauges with higher sensitivity have limited range. Of course, it's possible to purchase strain gauges with high capacity and sensitivity—for a price. If you check the Web for prices of high-capacity, high-sensitivity laboratory scales, you'll find that the dust covers for these scales cost more than our bathroom scale.

The sensitivity of a strain gauge is expressed in terms of Gauge Factor (GF). Mathematically, GF is expressed like so:

$$GF = (\Delta R/R)/(\Delta L/L)$$

Simply put, GF is the ratio of the fractional change in electrical resistance to the fractional change in length of a strain gauge. The fractional change in length, $\Delta L/L$, is the formal definition of strain. The greater the GF, the greater the sensitivity of a strain gauge. As a point of reference, the strain gauges in our scale have a GF of about 2.

If we're interested in measuring the weight of butterfly wings, given enough amplification, we could use the strain gauges in the current load cells for the job. However, given the stiffness of the metal supports, the amplification required would be so great that the signal-to-noise ratio would likely be too low to be useful without significant signal processing. A better approach might be to mount the strain gauges on a more flexible base, or to use a more sensitive strain gauge on a more flexible base.

The best solution depends on the requirements of the scale, including the operating environment. Scale-based strain gauges with a GF of 1000 might be perfect to measure the weight of butterfly wings—assuming the scale is in a lab free of vibration from nearby road traffic, and the scale isn't buffeted by air from a nearby air-conditioning vent.

Although I've focused on sensitivity in this treatment of strain gauges, I'd be amiss not to point out that strain gauges are fundamentally variable resistors. Moreover, the specifications typically associated with fixed and variable resistors apply to strain gauges, including tolerance, repeatability, resistance range, accuracy, linearity, thermal coefficient, power dissipation, operating temperature and humidity, and maximum applied voltage.

Of these variables, the most important from the perspective of an inexpensive bathroom scale is accuracy. The accuracy of a load cell is usually expressed as a percentage of full-scale load, with typical values ranging from 0.03 to about 1 percent of full scale. Assuming our 330-pound capacity scale uses 1 percent load cells, the value displayed on the LCD should be accurate to within ±3.3 pounds. I'm assuming an accuracy of 1 percent, given the price of the scale.

As noted in the discussion of specifications, the manufacturer does not provide an accuracy figure. If the accuracy is in fact 1 percent of full scale, then displaying weight to a tenth of a pound is at best misleading. A display with greater precision doesn't improve the accuracy of the displayed value.

Mods

The easiest mods for this scale are to enhance the triggering circuit and extend measurements beyond the confines of the unit. My favorite source for strain gauges and load cells is Omega Engineering (www.Omega.com). The National Instruments Developer Zone (http://zone.ni.com) is the best source for information on strain gauge design and use.

Enhanced Triggering

Try substituting an outboard microphone and preamp for the piezoelectric transducer to turn on the scale with a hand clap or whistle. Adjust the sensitivity or frequency response of the enhanced trigger so that ordinary room noise doesn't activate the scale, resulting in prematurely depleted batteries.

Create External Load Cells

Although it's a challenge, you can create new load cells by attaching the foil strain gauges to other surfaces. You'll need a steady hand and a sharp blade to peel the strain gauges from their metal mounts. Use superglue (cyanoacrylate) to attach the strain gauges to new platforms that provide both expansion and compression. Try mounts made of thin, relatively flexible metal or plastic to increase the sensitivity of the scale.

You can also use new load cells with different accuracies and sensitivities, as long as the total resistance of each strain gauge is 2KΩ. If you use load cells with a greater equivalent GF than that of the current load cells, the weight displayed on the scale's LCD will be greater than the actual weight. The caveat, of course, is that you'll have to create a lookup table to convert the displayed weight to actual weight.

Chapter 4

Surge Protective Devices

Surge suppressors and other surge protective devices (SPDs), virtually unheard of before the advent of the PC, are one of the most popular electronics accessories sold. I own at least a half dozen surge suppressors, such as the one shown in Figure 4-1, a pair of power conditioners, and an uninterruptible power supply (UPS).

Why this mix of power technologies? Why spend hundreds of dollars on a power conditioner with fancy meters and lights and-or a UPS capable of an hour of battery backup power instead of a simple surge suppresser? And why not simply use an inexpensive power strip? The reason is that wall socket power can be dirty and fickle: A rare lightning strike can fry your laptop, citywide surges stress devices with linear power supplies, and spike generators in the form of home appliances can be lethal to audiovisual (AV) gear, game consoles, computers, receivers, and other sensitive AC-powered electronics.

FIGURE 4-1 Monster Power AV600 surge suppressor

The most common SPD is the surge suppressor power strip, which provides multiple outlets with active and/or passive electronics that respond to the spikes and, to a lesser degree, surges from the power line. These inexpensive devices, typified by the Monster AV600, offer modest protection against nanosecond-duration events (spikes) and voltage increases that last a second or more (surges).

The typical consumer-grade surge suppressor employs MOVs (metal oxide varistors) to divert and attenuate voltage spikes. MOVs are inexpensive, relatively fast acting, and capable of conducting several thousand amps for a few microseconds, but they also degrade when they absorb energy from a voltage spike. Some commercial-grade surge suppressors combine MOVs with gas tubes, which have a higher energy capacity than MOVs, and avalanche diodes, which are faster than MOVs, to provide a mix of transient response speed, energy handling capacity, and longevity.

Another common SPD, especially in music and professional audio circles, is the power conditioner, which is often marketed as a significant improvement over a basic surge suppressor. Some power conditioners use transformers to provide isolation between equipment and the line and between components, heavy-duty inductors to filter noise from the hot and neutral wires, and a mix of nonlinear components to protect equipment from both surges and spikes. However, many power conditioners are little more than ordinary MOV-based surge suppressors in a fancy enclosure.

In this chapter, we'll tear down two SPDs. The first, an AV600 surge suppressor by Monster Cable, is available on the Web for about $14. The second device is a Furman PL-Plus C power conditioner, available for about $220. The teardowns of these Chinese-built units provide a practical introduction to MOVs, thermal fuses, thermal circuit breakers, component failure mechanisms, and electrical standards.

Surge Suppressors

Surge suppressors are commodity items. Inexpensive surge suppressors from Belkin, Tripp Lite, APC, or other manufacturers typically offer the same suppression features but vary in relatively insignificant features, such as the number and layout of sockets, color, housing size and shape, and length of power cord. I selected the Monster Cable AV600 surge suppressor for this first teardown because it's inexpensive and readily available, and the specifications are representative of similarly priced devices on the market.

Highlights

The AV600 is a typical MOV-based surge suppressor strip constructed of components designed to respond to temperature and heat—properties that aren't easily conveyed by conventional electronic schematics. We'll explore how surges and spikes are handled by separate circuitry and the significance of the "Protection On" LED (light-emitting diode) indicator. During this first teardown, note the following:

- The fiberglass wrap enveloping some of the components
- The slots in the circuit board
- The physical proximity of components
- The size, mass, and number of MOVs

Specifications

According to Monster Cable (www.monstercable.com), the six-outlet unit provides the following:

- Protection against both power surges and voltage spikes
- Suppression from hot to ground, hot to neutral, and ground to neutral
- 555 joule (J) energy rating
- 1 nanosecond (ns) response time
- 15A maximum load capacity at 120VAC, 60Hz
- 330V transient suppression voltage
- Compliance with UL 1449 standards

Let's consider the key specifications in more detail.

Joule Rating

The spike-suppression function of the AV600 works by diverting the spike energy from hot and neutral to ground, and the MOVs that perform this function have a finite energy-handling capacity, which is reflected in the joule rating. The *joule* is a unit of energy, with 1 joule equivalent to 1 watt-second. The greater the surge suppressor's joule rating, the greater its capacity to divert unwanted energy away from sensitive equipment.

Ideally, a surge suppressor can divert an infinite amount of energy from hot and neutral to ground—think lightning strike. However, despite marketing claims, surge suppressors are useless if your house wiring suffers a direct lightning strike. The same goes for a $900 power conditioner or UPSs. Your best bet is to unplug—not simply turn off—your equipment and surge suppressors during a lightning storm.

The energy rating of 555J is the sum of the energy-diverting capabilities of the MOVs used in the AV600. It's difficult to determine the specific energy-diverting capabilities of the AV600 without knowing the number and rating of MOVs installed from hot to ground, hot to neutral, and neutral to ground. And we'll know this after the teardown.

Response Time

Voltage spikes are by definition brief. For a surge suppressor to be useful, it must be fast enough to divert an energy spike before it reaches sensitive equipment. A 1ns response time is excellent for surge suppressors based on MOVs. Even so, 1ns is slow enough that some of the energy of a spike will pass through the suppressor before the MOVs begin to conduct. For example, let's say a high energy spike has a duration of 2ns and that 80 percent of the energy is dispersed in the first nanosecond. Even if the MOV conducts at 100 percent capacity at 1ns, 80 percent of the energy spike gets through the suppressor. In reality, the MOV only begins to conduct at 1ns, meaning that perhaps 90 percent of the spike makes it through the surge suppressor without being attenuated or diverted.

An alternative to the MOV is the gas discharge tube, which lags behind MOVs in response time but has a much greater energy-handling capacity. Avalanche diodes, packaged as transient voltage suppression (TVS) diodes, respond much faster than MOVs but can't handle as much energy.

Transient Suppression Voltage

The TSV is the voltage at which the energy diversion of a MOV kicks in. Given the tolerances of MOVs and other components, the normal variability of the line voltage, and the degradation of MOVs with use, the TSV should be significantly greater than the nominal line voltage. If the TSV is too high, however, the MOVs will be ineffective in protecting equipment. A TSV of 330VAC is a good compromise for an inexpensive surge suppressor.

Suppression LG/LN/GN

Most MOV-based surge suppressors provide energy diversion between hot and ground (LG), hot and neutral (LN), and ground and neutral (GN). The cheapest MOV-based surge suppressor strips often have a single MOV between hot and ground, and therefore can't divert spikes from neutral to ground, clamp voltage spikes between hot and neutral, or divert spikes common to both hot and neutral (common-mode spikes) to ground.

Spike diversion is but one of several methods of protecting equipment from voltage spikes. I'm partial to "heavy iron"—large toroidal transformers that provide complete isolation from the mains, large inductors in the hot and neutral leads that attenuate noise and transients, and nonlinear devices that clamp any residual spikes. But all this comes at a significant price—which is why inexpensive surge suppressors based on 75-cent MOVs are so popular.

UL 1449

The UL 1449 rating is important because it certifies that the AV600 is safe to use. It doesn't certify that the AV600 will protect your equipment from a direct lightning strike or even a minor surge, however. According to the Underwriters Laboratories (www.ul.com), the 1449 rating does not evaluate the effect of the surge suppressor on connected loads, establish a minimum level of attenuation, or assess the adequacy of the voltage protection rating. In short, UL 1449 certification is assurance that the AV600 won't catch fire under normal use.

The UL 1449 standard undergoes periodic updates, and each new edition is more detailed than the last in the assessment of SPDs, including MOVs. UL 1449, third edition, became effective in September 2009. Several UL ratings are relevant to SPDs, depending on the design. For example, UL1414 deals with capacitors connected directly across or in series with the mains. International standards organizations, such as the IEC (International Electrotechnical Commission, www.iec.ch), also establish standards for SPDs.

Operation

I use my AV600s as central power control stations. After all, why flip on a half-dozen on–off switches when one on the surge suppressor will do? Simply hit the illuminated rocker switch and, if both the switch and "Protection On" LED are illuminated, you're good to go.

Before the teardown, I assumed the LED was a ground indicator. After all, a surge suppressor is supposed to divert energy to ground—the green wire, attached to the round conductor on the standard 3-pin NEMA (National Electrical Manufacturers Association) plug. However, in preparation for this teardown, I installed a 2-to-3-prong adapter on the AV600 cord and the "Protection On" LED remained illuminated. I'll discuss the LED circuit in detail later, but you should keep it in mind during the teardown.

Teardown

If you're following along at home with an AV600, you should be able to complete the basic teardown, illustrated in Figure 4-2, in about 5 minutes. If you want to disassemble the main components and remove the MOVs from the circuit board, then give yourself another 20 minutes. Note that this is a destructive teardown. We'll have to destroy the case to get at the components.

FIGURE 4-2 Teardown sequence

g

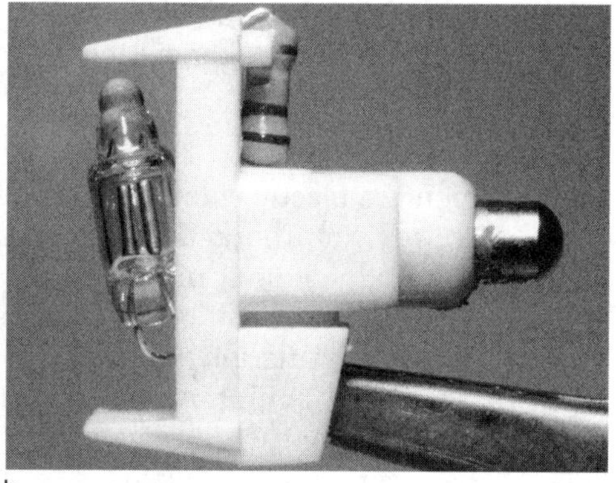
h

FIGURE 4-2 (*continued*) Teardown sequence

Tools and Instruments

For the basic teardown, you'll need a standard screwdriver with a large blade, a pair of gloves, and eye protection. A large pair of diagonal cutters is useful for extracting the circuit board from the plastic housing. You'll need a soldering iron, pliers, solder wick, or equivalent desoldering tools to extract the components from the circuit board. A drill with a 1/4-inch bit and a pair of heavy-duty lineman's pliers are required for the optional component teardowns.

You'll need a multimeter to test the state of the fuses and other components in the AV600. If you want to experiment with the MOVs, you'll need a high-voltage source, some means of generating voltage spikes, and an oscilloscope. I use a battery-powered photoflash unit to generate spikes. A Variac variable transformer is a great tool for creating surges.

Step by Step

As you progress with this short teardown, you may find it useful to tape down the power cord and other wires as your free components, in preparation for testing.

Before you begin the teardown, verify that the unit is unplugged.

Step 1

Examine the exterior of the unit. There is a 6-foot, three-conductor, 14AWG (American Wire Gauge) power cord attached to one end of the surge suppressor. Note the circuit breaker reset button on the side of the unit, about 1 inch from the cord end of the unit, as shown in Figure 4-2a. Also note the "Protection On" LED, just above the rocker switch.

Step 2

Open the plastic clamshell. With gloved hands, direct the flat blade of a large screwdriver around the unit, breaking the lightly glued seam. Next, insert the blade into the body of the suppressor, around the midline. Pry the halves apart to reveal the circuit board and resettable fuse, as shown in Figure 4-2b. Note the damaged insulation (as shown in the figure) on the ground and hot wires where they enter the unit. Apparently the power cord was not seated properly when the two halves of the clamshell were assembled—an apparent lapse in quality control and a potential fire hazard.

Step 3

Extract the circuit board, LED, and circuit breaker. I removed the board and switch assembly by grasping the plastic body with gloved hands and snapping the body in half. If you're following along at home and want a more controlled approach, use diagonal cutters to nibble away at the plastic body until you free the circuit board. When you're done, you should see the power cord, circuit breaker, circuit board, LED, and the leads to the power jacks, as shown in Figure 4-2c.

To make the photo in Figure 4-2c, I removed a white fiberglass cloth covering the four dark rectangular components (MOVs) to the right of the switch, and a second cloth covering the vertically mounted 1W resistor and the white rectangular component (thermal fuse) near the midline of the board. The separate fiberglass cloths insulate and connect the components thermally and provide protection from fire should a component ignite.

Note how the ground wire from the power cord is soldered to the ground wire of the ground bus, and that the neutral and hot wires are welded to their buses. Note also that the hot (black) wire is the only one that's interrupted by the circuit board. The ground and neutral wires make contact with the board, but they continue on to the receptacle bus. The figure also shows the damaged ground wire insulation.

Step 4

Examine the foil side of the circuit board, as shown in Figure 4-2d. Note the relatively wide circuit tracks, the wide spaces between traces, and the air gaps throughout the middle and right side of the board. Any idea why the manufacturer went through the expense of creating the air gaps? Have you ever seen a circuit board charred by a lightning strike?

Step 5

Identify the components on the board. Unfold the white rectangular thermal fuse adjacent to the 1W resistor, shown in Figure 4-2e. As indicated by the package markings, this is a 115°C thermal fuse with a rating of 15A at 250VAC. Next, examine the double-decker sandwich of four dark rectangular MOVs and two cylindrical thermal fuses, shown in Figure 4-2f. Note the "102°C" marking on the thermal fuse, as well as the markings on the MOVs.

Step 6

Disassemble the thermal circuit breaker (optional). Drill out the five rivets holding the two halves of the component together and use a pair of pliers to remove the sleeve around the push button. You don't have to drill through the casing of the circuit breaker—simply cut through one end of a rivet to release it. Examine the contact mechanism, shown in Figure 4-2g.

Step 7

Disassemble the rocker switch (optional). The rocker switch contains a neon bulb worth studying. If you haven't worked with a neon bulb before, now's the time to harvest one for your parts collection. Simply cut through the soft plastic housing with diagonal cutters to free the bulb and series resistor assembly, as shown in Figure 4-2h.

Layout

As revealed in the teardown, the majority of components are located on a compact circuit board about 1 1/2 × 1 3/4 inches (38mm × 45mm), shown in Figure 4-3. The board contains the illuminated rocker switch, four rectangular MOVs, three

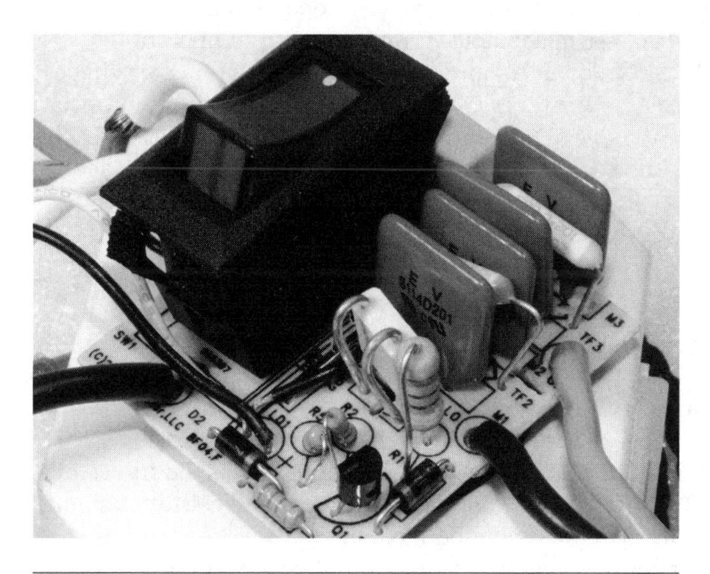

FIGURE 4-3 Circuit board layout

thermal fuses, a transistor, a pair of general-purpose silicon diodes, a capacitor, and a few resistors. The thermal circuit breaker and two of the three power buses—hot and neutral—make connections with the circuit board.

Components

Initially, you might be surprised at the number of components inside such a simple device. After circuit analysis, however, you might be equally impressed at the number of functions these few components can accomplish.

Thermal Circuit Breaker

The largest single component and the only major component not soldered directly to the circuit board is the Joemex (www.joemex.com) thermal circuit breaker, visible in Figure 4-2c. The push-to-reset circuit breaker, rated at 15A and 120VAC, is wired in series with the hot wire, between the power cord and the circuit board.

The circuit breaker consists of a bimetallic band fixed on one end with a contact on the free end, as shown in Figure 4-2g. When the temperature of the strip reaches a predetermined level, whether due to current through the band for a given time or the ambient temperature, the metal band bends, thereby opening the contact. After the band cools for a minute, the plunger on the button can be used to force the contacts together, setting up the circuit breaker for the next over-current event.

The bimetallic arm is a temperature- and time-dependent device. At room temperature, the circuit breaker can carry 15A indefinitely. With a 20A load at room temperature, the circuit breaker will open in about 1 hour. A 30A load will trip the circuit breaker in about 40 seconds at room temperature.

As the ambient temperature rises, the current carrying capacity of the breaker diminishes. For example, at 86°F, the current carrying capacity of the circuit breaker is decreased by 10 percent. At 104°F, a temperature easily achieved in a closed computer closet or AV cabinet, capacity is decreased by about 30 percent.

MOVs

The four EV S114D201 MOVs in the AV600 are rectangular, epoxy-dipped components resembling ceramic disc capacitors. At their normal working voltage of 120VAC, the MOVs are essentially capacitors with a low leakage resistance. I measured the average capacitance of the four MOVs as 800pF and a DC resistance in excess of 10MΩ.

When the voltage across a MOV approaches the transient suppression voltage of 330VAC, the MOV behaves more like a low-resistance shunt, both absorbing and diverting energy. The amount of current that can be carried by a MOV is limited. The higher the current, the higher the heat generated in the body of the MOV, and the greater the likelihood of permanent damage. Recall that the heat produced by a resistance is proportional to the square of current ($P = I^2R$). Unlike resistors, MOVs aren't designed for continuous dissipation of significant energy, but only brief surges.

The 330VAC TSV is based on the specifications of the AV600 unit. Although I was unable to locate the MOVs in a catalog, the component markings suggest a 14mm square body (S114), normal energy rating (D), and a varistor voltage of 200VDC (20×10^1). The varistor voltage, the DC voltage at which 1mA current flows, is one means of identifying the voltage-handling capabilities of a MOV. The varistor voltage is typically, but not necessarily, less than the maximum clamping voltage.

Typical catalog listings for MOVs include the maximum continuous AC and DC voltage, the single pulse transient energy capacity, the varistor voltage, the transient suppression voltage, and capacitance. The transient suppression voltage rating of a MOV is proportional to the thickness of the device, current capacity is proportional to surface area, and energy absorption capacity is proportional to volume (that is, thickness × surface area). The MOVs in the AV600 are 0.5 inch square × 0.06 inch thick (14.6mm sq × 2.1mm).

I set up the experiment illustrated in Figure 4-4 to verify my reading of the component varistor voltage. I used a linear high-voltage power supply from a vacuum tube amplifier to apply a constant DC voltage across a MOV while monitoring the current. Below 200VDC, there was essentially no current flow. However, at 213VDC, the current was 1mA, indicating a varistor voltage of 213VDC. Given typical component tolerance and instrument accuracy, this reading is consistent with a varistor voltage of 200VDC, as per the component marking. Based on varistors listed by DigiKey, a varistor voltage of 200VDC is compatible with the 330VAC transient suppression voltage rating.

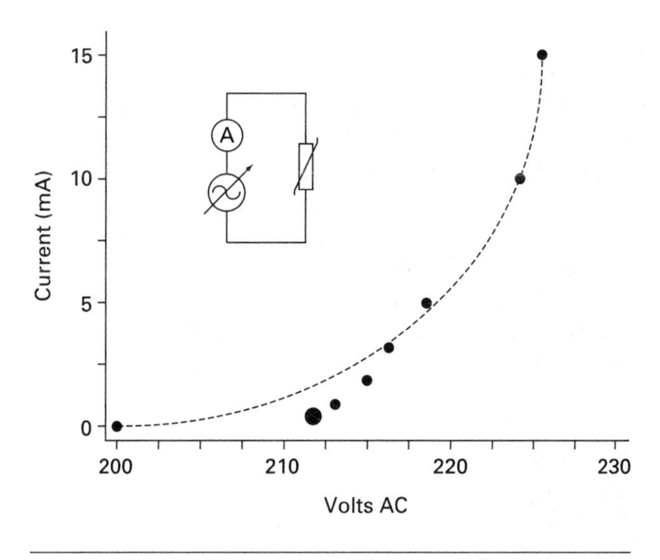

FIGURE 4-4 MOV voltage-current curve experiment results

A limitation of MOVs is that they deteriorate with use. In response to each spike they encounter, a sudden increase in current raises the temperature of the MOV, decreasing its resistance, and, assuming the spike hasn't abated, increasing current. Unless the cycle is checked, it will continue until the MOV fails, typically as a short. In the case of a low-resistance short, the MOV may explode or flame prior to tripping the circuit breaker. This is one reason for the fiberglass wrap; if a MOV fails catastrophically, the heat and shrapnel will be contained within the fiberglass.

If a MOV fails to resistance of, say, 100Ω, the temperature of the component will increase, elevating the temperature of nearby components. Eventually, the MOV will melt, crack, and, ideally, become an open circuit without flaming. Obviously, there's a place for an automatic means of detecting an abnormal rise in MOV temperature and for disconnecting it from the circuit before a damaged MOV becomes a fire hazard.

Because the condition of the MOVs is not apparent through the AV600's opaque case, an indicator of MOV status would be useful. Although it's not practical in an inexpensive surge suppressor, one means of assessing the health of a MOV is to monitor its varistor voltage. It's generally accepted that damage has occurred when the MOV's varistor voltage changes by 10 percent or more from its initial value.

Thermal Fuses

A thermal fuse, also known as a thermal cutoff, permanently opens when its temperature rises above a temperature set point. As with regular current-based fuses, thermal fuses also carry a current rating. However, this figure represents the maximum operating current that a thermal fuse can withstand without derating its temperature set point.

Consider the fuse sandwiched between the two MOVs in Figure 4-2f, for example, which is rated at 5A and 102°C. When the temperature exceeds 216°F (102°C), the fuse permanently opens. If the temperature is only 180°F, the fuse won't blow at a current of 5A or even 10A. The set temperature may decrease from 216°F to 197°F at 10A, but otherwise, within the current carrying capacity of the fuse, it won't open. In my tests, the 102°C/5A thermal fuses in the AV600 can withstand 20A for 1 minute at room temperature before failing. I'm not suggesting you ignore the current rating of a thermal fuse, but just don't confuse the rating with that of a standard fuse. Using a 102°C/5A thermal fuse in a 10A circuit is akin to using a 5A/250VAC slow-blow fuse at 400VAC.

In addition to the thermal fuses sandwiched between MOVs, is a larger, separate thermal fuse, shown in Figure 4-2e, with ratings of 115°C (239°F) and 15A. As with the smaller fuses, the 15A fuse can carry significantly more current than 15A, at the cost of shifting the temperature set point. As you'll see later in the circuit analysis, this thermal fuse provides protection from long-duration voltage surges.

Rocker Switch Neon Lamp

The neon lamp in the rocker switch is worth examining because it represents a class of lamps that is used as voltage regulators and voltage spike diverters. As shown in Figure 4-2h, the neon lamp is constructed of two cylindrical electrodes enclosed in a glass envelope that is filled with neon and a trace of argon under low pressure. When AC or DC voltage is applied to the electrodes, the gas ionizes and gives off an orange-yellow glow. With AC, the area around both electrodes is ionized. With DC voltage, only the area around one electrode is ionized. Do you know which one?

As switch indicators, neon bulbs are useful in line-powered circuits because they're bright, use little power, and require only a series current-limiting resistor to operate at 120VAC. The bulb in the AV600, which is about a third smaller than the ubiquitous NE-2 neon bulb, is connected to a 150KΩ series resistor. This results in an operating current of 120VAC/150KΩ, or 0.8mA.

Unlike an incandescent bulb, the neon bulb exhibits a hysteresis (delayed) effect. The breakdown voltage is about 70VAC/DC, but once the gas in the bulb is ionized, it takes about 10V less voltage to maintain ionization. Another feature of the neon bulb is that the voltage drop across an ionized bulb is constant. Neon bulbs with a breakdown voltage of 200VAC or more are used as spike arrestors in power supplies, power conditioners, and tube-type radio receivers. Neon bulbs are relatively slow to respond to spikes, but their instantaneous shunt current capacity is significant.

Transistor

The single transistor in the AV600 is a KSP44, a high-voltage silicon transistor in a TO-92 package, visible in Figure 4-3. Following the standard package format, looking at the flat face of the transistor, the leads, from left to right, are emitter, base, and collector. The transistor is rated at 500V collector-base and 400V emitter-base voltages, with a collector current of 300mA. The datasheet is available through the Fairchild semiconductor web site (www.fairchildsemi.com).

Diodes

The two diodes in the AV600 are generic silicon power diodes, type 1N4007. They're rated at 1000V PIV (peak inverse voltage) and 1A forward current.

Capacitor

The single, yet prominent, capacitor in the circuit is an MKP X2 metalized polypropylene film capacitor (www.dain.com.tw) rated at 0.1μf and 250VAC. The non-inductive capacitor, constructed with a plastic case sealed with epoxy resin, is sold as an interference-suppression capacitor, for applications that include antenna coupling, bypassing, and electromagnetic interference (EMI) filtering.

Resistors

Of the four metal film resistors in the AV600, the most notable is the 1W, 880Ω, fusible resistor that's thermally isolated with the 15A thermal fuse, as shown in Figure 4-3. The resistor has a flame retardant coating that enables it to fail gracefully, without creating a fire hazard. In normal operation, a band of fiberglass cloth snugly binds the large thermal fuse to the resistor.

LED

The single red LED serves as the "Protection On" indicator light. However, as you'll see in the teardown, what the LED signifies is probably not what you expect.

Circuit Board

The single-sided circuit board is remarkable in the seven cutouts between conductive islands, visible in Figure 4-2d. The cutouts are air-gap insulation between components and leads of the same component that could be compromised by a high-voltage spike. One reason for the gaps is to avoid charring of the circuit board.

Cord

The cord and connector are integral to the AV600. In order for the MOVs to work properly, the unit must be grounded—hence the standard three-pronged NEMA plug. The 6-foot power cord consisting of stranded 14AWG copper wire is sufficient for supporting a 15A load. As a point of reference, 14AWG copper wire has a resistance of about 0.0025Ω/foot at room temperature.

How It Works

The typical single-phase residential power drop-off consists of a hot (black wire), neutral (white wire), and ground (green wire). The hot and neutral lines connect to the power distribution system and the ground is derived locally by connecting a copper wire to a local earth ground, such as a cold water pipe. The flow of current from the hot or neutral to the ground lead—a ground fault—is an abnormal condition. However, the ground lead is a handy place to divert the unwanted energy of a voltage spike, sparing your expensive electronic gadgets.

With this brief background, let's look at the schematic for the AV600, shown in Figure 4-5. Input is on the left and output is on the right, represented by the two NEMA jacks. Note the TO-92 package outline for the PNP transistor (Q1).

The hot lead is interrupted, from left to right in the schematic, by the resettable 15A thermal circuit breaker, by the power switch assembly, and by the 115°C thermal fuse. The symbol for a thermal fuse reflects the typical packaging—a hollow metal tube filled with material from the open end. The overflow of material on the open end tapers to the lead, forming the point of the package. The thermal fuses in the AV600 don't happen to use standard packaging. The other schematic symbol that may be new to you is a rectangle with a diagonal line through its body, which represents a MOV.

The neon bulb (NE) and 150KΩ series resistor, between hot and neutral, illuminate the translucent rocker switch (SW) when the rocker switch is in the on position, assuming the 15A circuit breaker is functional. The 0.1µf capacitor across the rocker switch contacts suppresses the arcing that might occur if a significant load is attached to the surge protector when the switch is moved on or off. Without the capacitor, the arcing across the contacts could damage the equipment plugged in to the AV600 as well as the rocker switch.

Now let's step back and look at the AV600 from a systems' perspective. There are five functional subsystems: a current-limiting system, a spike-diversion system, a spike-clamping system, a power-indicator system, and a surge-protection system. The current-limiting system consists of the 15A thermal circuit breaker. If the load exceeds 15A, the breaker will open, with a response time dependent on the load. A dead short will trip the breaker in a second or two.

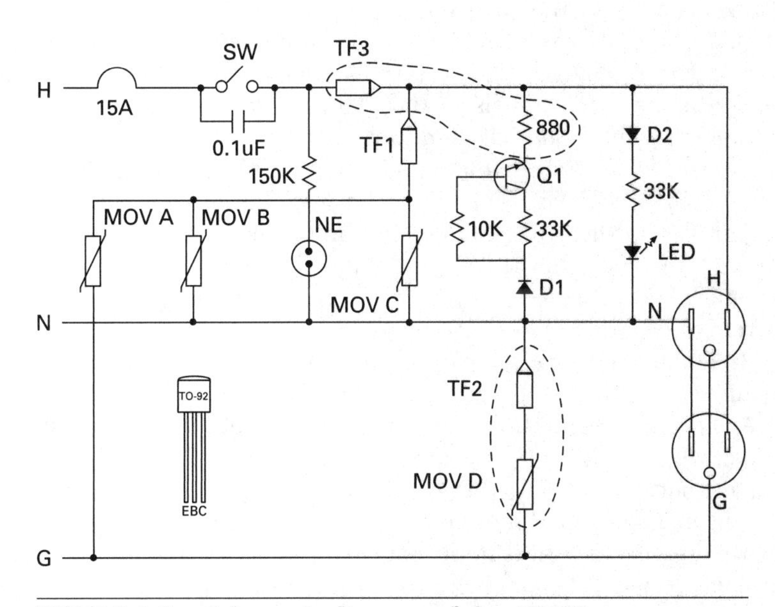

FIGURE 4-5 Schematic diagram of the AV600

The spike-diversion system consists of MOV A between hot and ground and MOV D between hot and neutral. A voltage spike on the hot or neutral lines will be diverted to the ground wire. Note that if the ground wire isn't grounded, the MOVs will be ineffective.

The spike-clamping system consists of MOVs B and C that will clamp a voltage spike between hot and neutral. In compliance with UL 1449, the MOVs are not placed directly across the lines, but are wired in series with thermal fuses. Moreover, the two 102°C thermal fuses (TF1 and TF2) are physically and thermally associated with the MOVs they protect. The dotted ellipse enveloping the TF2 and MOV D signifies one of the thermal links. If MOV A absorbs a spike and fails with a short, its temperature will quickly reach 216°F (102°C) and TF1 will open, removing both MOV A and MOV B from the circuit. This alleviates a potential meltdown of the AV600 but also diminishes the protection provided by the unit.

The power-indicator system consists of the neon bulb within the power switch and the "Protection On" LED between hot and neutral. As long as TF3 is functional, flipping the power switch to the on position should illuminate both neon bulb (NE) and LED. If TF3 is open, the rocker switch will be illuminated, but not the LED. In other words, the state of the LED reflects the status of TF3, independent of the status of the MOVs or other thermal fuses.

The surge-suppression system consists of TG3, a 115°C thermal fuse that is thermally and physically associated with the 880Ω, 1W resistor. The dotted loop enclosing both components at the top of the schematic highlights the thermal connection. If you recall from the teardown, the 115°C thermal fuse and 880Ω resistor are wrapped in an insulating fiberglass cloth.

The 880Ω resistor functions as a thermal indicator of the voltage between hot and neutral. As the voltage increases above 120VAC, the power dissipated by the resistor increases and its temperature rises. The temperature of TF3 tracks the temperature of the resistor. If the sufficiently elevated line voltage persists, the temperature of the resistor and TF3 will reach 239°F (115°C), and TF3 will open. So why the diode (D1) and transistor (Q1)?

Let's examine Figure 4-6 to understand the design of the surge-suppression circuit. The figure shows a plot of voltage versus current and the surface temperature of the 880Ω resistor. The first thing to note is the small line segment labeled "1mA/Volt" in the lower-right quadrant of the graph. It represents the slope of voltage versus current for pure resistance plotted against the same voltage and current scales used in the larger graph.

In comparison, the slope of the voltage versus current for the resistor-transistor circuit combination starts out at about 0.5mA/Volt and then increases to more than 2mA/Volt. As a result, the surface temperature of the 880Ω resistor, and the body temperature of the 115°C thermal fuse, increase rapidly with an applied voltage greater than 120VAC. In other words, with increasing voltage, the relative increase in current increases—and more rapidly than it would with a resistor alone. The purpose of the transistor circuit is to introduce nonlinearity in the flow of current through the 880Ω resistor, thereby increasing the sensitivity of the surge-suppression system to prolonged overvoltages.

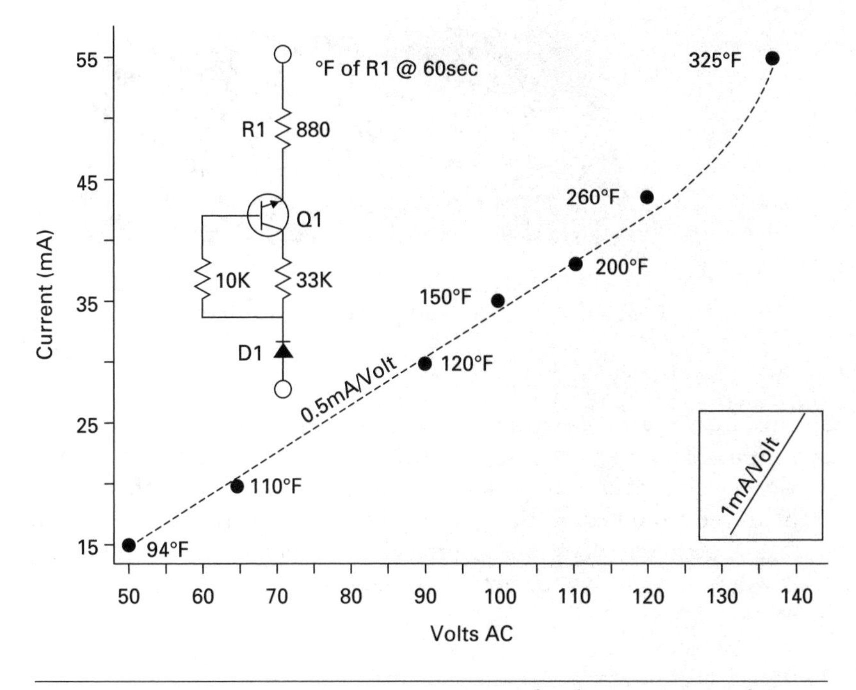

FIGURE 4-6 Surge-suppression circuit and voltage-current plot

The D1 ensures that the transistor conducts only when the hot is negative relative to the neutral wire. The two resistors provide forward bias for the transistor. The values of the 10 and 33KΩ resistors were apparently chosen to position the Q1 in a particular region of its operating curve so that a significant nonlinear change in the voltage-current relationship occurs when the emitter-collector current is about 45mA. See the datasheet for details on the operating characteristics of the KSP44s transistor.

Mods

I don't suggest modding the AV600 for regular service, because of the potential fire hazard. However, as a bench experiment, a worthy goal is implementing a more useful "Protection On" indicator that reflects ground and MOV status. Perhaps you can sell the circuit design to Monster Cable.

Power Conditioner

The subject of this supplemental teardown is the Furman PL-Plus C Power Conditioner, hereafter referred to as simply PL-Plus and shown with its top cover

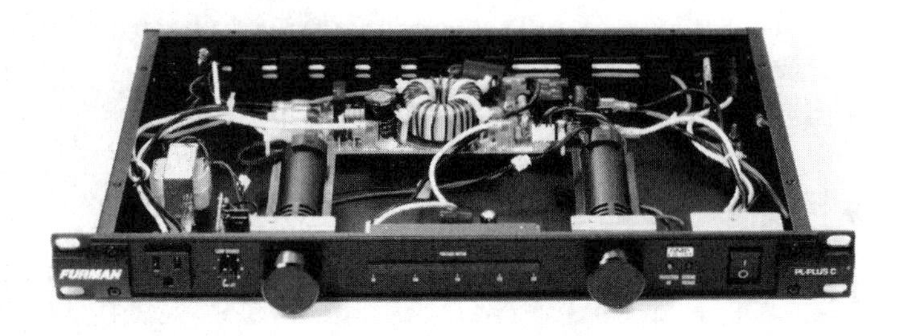

FIGURE 4-7 Furman power conditioner

removed in Figure 4-7. I selected the rackmount Pl-Plus over similarly priced power conditioners from Monster Cable, Tripp Lite, and others for this teardown simply because I own several previous models of Furman conditioners and they've been trouble-free from installation day. Part of the attraction of the PL-Plus is its user interface. Not only does it offer nine outlets, including one in front, but there's an LED voltmeter; pullout, dimmable LED lights; and a BNC socket in the rear for a gooseneck lamp.

Highlights

Given its cost, I assumed the PL-Plus was a super surge suppressor with the added benefits of lights and built-in voltmeter. However, as you'll see in the teardown, I didn't find quite what I expected. Given our previous discussion of surge suppressors, note the following:

- MOV count, capacity, and placement
- Treatment of the ground wire
- Provision for MOV failure

Specifications

Based on the marketing literature, packaging, and markings on my PL-Plus, specifications for the nine-outlet unit include the following:

- Isolated banks of sockets
- 15A maximum load capacity at 120VAC, 60Hz
- 1ns maximum response time
- Protection against both power surges and voltage spikes
- 12W power consumption
- 133VAC RMS spike-clamping voltage and 188V peak at 3000A

- Maximum surge current of 6,500A
- Automatic shutdown with prolonged voltage over 140VAC and under 90VAC
- Spike suppression from hot to neutral
- Noise suppression 10dB at 10kHz, 40dB at 100kHz, 50dB at 500kHz
- Compliance with TÜVRheinland standards

Some of the specifications are unclear. For example, does the 12W power consumption refer to the underlying protection circuitry or simply the drain of the lighting system and LED voltmeter? How does the 3000A figure differ from the 6,500A maximum surge current? How are the banks of isolated sockets segregated, and what is the isolation in dB?

Also of note is what's not mentioned. There is no mention of Joule rating, for example. There is no mention of neutral-to-ground or hot-to-ground transient suppression, only that there is no ground contamination. There's also no mention of UL listing anywhere in the literature or on the body of the unit. Furthermore, a visit to the TÜVRheinland web site (www.tuvdotcom.com) failed to provide me with much insight into the TÜV standard.

In the realm of curious statements, the user manual states that, unlike traditional surge-suppression circuits that sacrifice themselves when exposed to multiple transient spikes, requiring repair, this isn't the case with the PL-Plus. The unit somehow protects itself. Furman also states that MOVs, when used within their design specifications, do not degrade, and that the circuitry within the PL-Plus contains both a 12,000A MOV and a 12,000A TVZ (thermo-fuse varistor). Furthermore, an overvoltage protection circuit protects the MOV from damage caused by overvoltage. Other claims include the shunting of transients only 1.5 to 2.0V above the peak line voltage to an internal capacitor, and a filter circuit that attenuates a typical lightning voltage surge.

Given all of these claims, most of which require additional information for clarification, I was curious about what I'd find inside the PL-Plus. Let's just say I was surprised.

Operation

There isn't much to say on the issue of operation. You plug in your equipment, either in the front or rear sockets, and flip the front on–off rocker switch. If all is well, the LED voltmeter should show the current line voltage and the green "Protection On" LED in the front panel should be illuminated.

If you mistakenly plug the unit into a 240VAC line—hard to do with a NEMA 15 plug—the red "Extreme Voltage" LED will illuminate, the LED voltmeter will dim, and the hot lead to the sockets will be interrupted. The same goes for a brownout—if the line voltage drops below about 90VAC, the hot lead to the nine sockets will be interrupted. It's important to note that the neutral wire is not affected by the over- and under-voltage cutout circuitry.

Mini-Teardown

This is a mini-teardown, perhaps better described as a simple extraction, as illustrated in Figure 4-8. I'm going to focus on the components and circuitry that deal directly with voltage and surge suppression. If you're following along at home, give yourself 5 minutes to pop the top of the unit and another 5 minutes to extract the circuit board.

Tools and Instruments

You'll need a Phillips-head screwdriver with a sharp, small to medium head to get at the cover screws. Use a manual screwdriver—I found the screws easy to strip

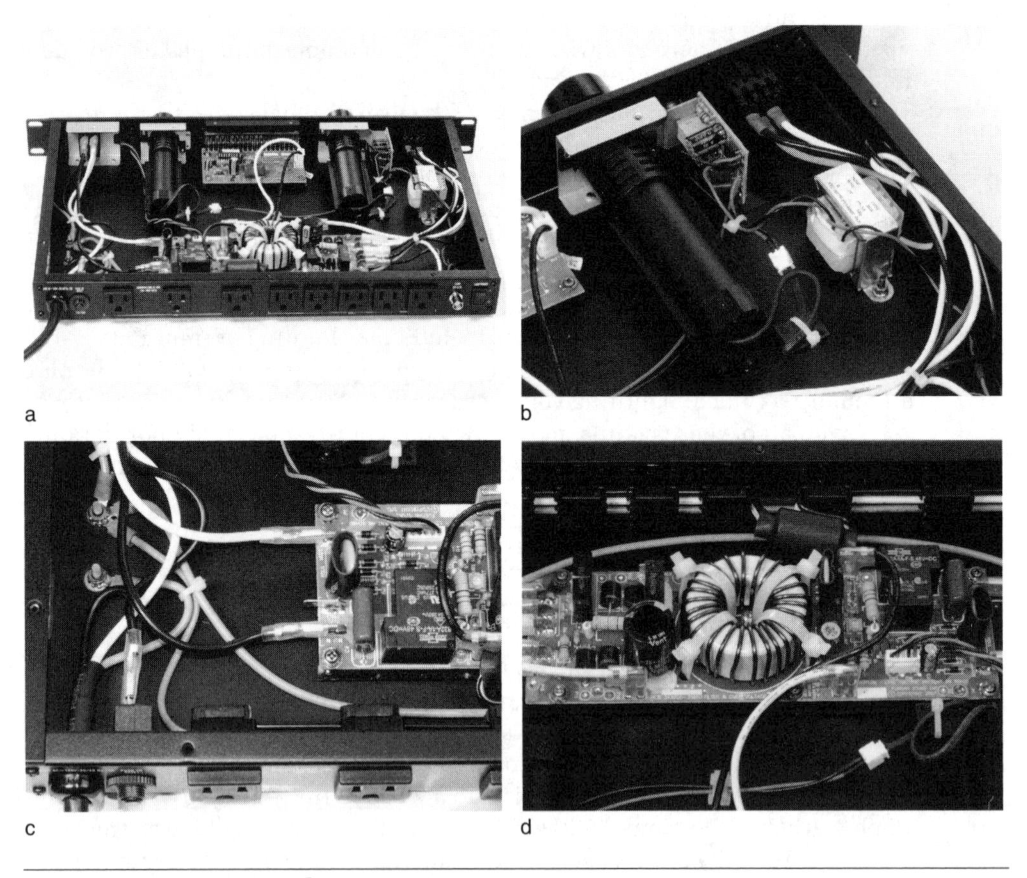

FIGURE 4-8 Mini-teardown sequence

with a power screwdriver. A pair of needle-nose pliers will be useful in breaking the quick-disconnect hardware connections. You'll need a Variac and multimeter to test the voltage cutoff circuitry.

Step by Step

Before you begin the teardown, verify that the unit is unplugged. It's also a good idea to remove any devices plugged into the unit.

Be sure the unit is unplugged before you begin the teardown.

Step 1

Remove the top plate and ground wire. Remove the screws securing the top plate. Next, carefully tilt the plate until you can see the ground cable connection. With a pair of pliers, unplug the ground wire and put the top plate to the side.

Step 2

Identify the main subsystems. Turn the unit so that you're facing the rear sockets, as shown in Figure 4-8a. On the outside of the unit, from left to right, locate the grommet securing the three-conductor, 14AWG power cord to the chassis, the 15A circuit breaker reset button, eight 3-prong NEMA 15 sockets, the 12V external lamp connector, and the oddly placed on–off switch for the external lamp.

Now, focusing on the inside of the chassis, identify the power circuit board, the largest and most prominent subsystem, located adjacent to the output sockets. On the far right is the 12V lamp circuit, including the 12VAC power transformer bolted to the bottom of the chassis and the small dimmer circuit board, shown in Figure 4-8b. Between the cylindrical slide-out LED lamps is the LED voltmeter circuit board, visible in Figure 4-8a. Note the ground wire connections to the chassis, shown in Figure 4-8c, and the lack of connectivity with the power circuit board.

Now, turn the unit so that you're facing the front panel. Note the connections to the power circuit board as well as the wiring of the rear sockets, shown in Figure 4-8d. Notice how the hot and neutral lines attach to the power circuit board on the right and exit on the left.

Hot, neutral, and ground wiring within the unit are 14AWG stranded copper, with the exception of the wire running through the Roto-Tech NEMA sockets in the rear of the unit, which is 14AWG solid copper. Also, note the shrink-wrapped RFI (radio frequency interference) inductor around the line and neutral wires, between two of the Roto-Tech sockets. Note how the insulation of the inductor presses against several components on the power circuit board.

Step 3

Extract the power circuit board. Note the location of the quick disconnect tabs as you remove the wires connected to the circuit board. Next, remove the seven Phillips-head screws securing the board to the bottom of the steel chassis. Flip the board over to examine the trace pattern. The neutral trace runs uninterrupted on one side of the board. The hot trace is along the opposite side of the board, interrupted at the location of the relay and of the toroidal inductor.

Layout

Let's focus on the layout of the 8 × 2 1/2 inch power circuit board, shown in Figures 4-9 and 4-10. As noted during the teardown, the board has connections to the hot and neutral wires. The input section of the board, shown in Figure 4-9, contains two MOVs, a thermal fuse, a relay, a rectangular metalized polyester film capacitor, and circuitry to drive the two status LEDs on the front panel. The output or post-inductor section of the board, shown in Figure 4-10, includes a bridge rectifier circuit, an electrolytic capacitor, a small inductor, and a pair of rectangular metalized polyester film capacitors.

Of note are the two MOVs in Figure 4-9. The MOV on the edge of the board stands alone, soldered across the line and neutral lines. However, the MOV nearest

FIGURE 4-9 Input section of the power circuit board

FIGURE 4-10 Output section of power circuit board

the large inductor is wired in series with the thermal fuse. I removed the heat-shrink tubing that physically bonded the thermal fuse and MOV for illustration purposes.

Components

The following discussion of components is limited to those on the power circuit board and mounted on the chassis that are involved in transient and surge suppression.

MOVs

Two Sincera MOVs are on the power circuit board, a 20D391K-N and a 20D201K-N. According to the Sincera documentation (www.worldproducts.com), the 20mm MOVs are standard models, with varistor voltages of 390VDC ($391 = 39 \times 10^1$) and 200VDC ($201 = 20 \times 10^1$). The *K* signifies 10 percent tolerance and the *N* signifies the component is lead-free. The phenolic coating on each MOV is rated at 2000VDC.

Surprisingly, given the specifications of the PL-Plus, high-energy MOVs are not used. In addition, the typical response time for the MOVs is listed as <15ns, as opposed to 1ns response time listed by the PL-Plus manufacturer. The MOV catalog also lists criteria for determining MOV failure—the standard voltage change of

±10 percent from the initially measured varistor voltage, an apparent admission that the MOVs deteriorate with use. More significant is the maximum clamping voltage, which the catalog shows as 650V for the 20D391K-N and 300V for the 20D201K-N. I measured the thickness of the MOVs at 2.5mm for the 200VDC unit and 2.9mm for the 390V unit, which is in agreement with the voltage rating.

Thermal Fuse

The single thermal fuse on the board, which has a trip temperature of 115°C, is physically bonded to the lower voltage MOV with shrink-wrap tubing, as shown in Figure 4-9. This thermal fuse has the traditional shape with one square end and one pointed end.

Thermal Circuit Breaker

The Rong Feng 125VAC/15A thermal circuit breaker is similar to the one discussed earlier. It's one of the few significant components not mounted on the power circuit board.

Inductors

The unit includes three inductors—the large toroidal choke in the center of the board, a small choke on the output section of the board, and a tubular ferrite choke around the hot and neutral leads between several of the rear sockets. The large choke, composed of 26 turns of enameled 12AWG copper wire wrapped on a 2-inch diameter and 1-inch thick core, accounts for the majority of the board's weight of 15 ounces (430g). I measured the inductance of the choke at 0.16 milliHenry (mH). This equates to an inductive reactance of about 1Ω/kHz, using the formula $XL = 2\pi fL$. At 60Hz, inductive reactance is only 0.001Ω, but at 100kHz, it climbs to a modest 100Ω.

The small 22μH (microHenry) inductor, about the size of the 1/8W film resistor, is on the post-inductor side of the power circuit board. The inductor is part of a series circuit with a resistor and capacitor.

I was unable to determine the inductance of the tubular ferrite choke, which apparently accounts for the manufacturer's claim of isolating the power sockets. Based on ferrite chokes of similar size listed in the Digi-Key (www.digikey.com) catalog, the choke probably introduces only about 50Ω of resistance to RF signals at 1MHz traveling in the hot and neutral lines.

LEDs

Two generic LEDs are located in the front of the unit, one red and one green. The green LED serves as the "Protection On" indicator, and the red the "Extreme Voltage" indicator.

Rocker Switch

The on–off rocker switch is a DPDT (double-pole, double-throw) switch that connects and disrupts both the hot or line and neutral wires. This model has no internal light.

Diodes

Four 6A4 silicon diodes reside on the output side of the power circuit board, shown in Figure 4-10. The 9mm diameter diodes are rated at 400V PIV with a maximum peak forward surge current of 400A. The diodes are part of an unusual bridge circuit on the post-inductor side of the board.

Capacitors

Three 0.47µf at 275VAC metalized polyester film capacitors are packaged in rectangular, flame-retardant plastic cases and noncombustible resin. According to the datasheet, these capacitors are typically used in interference suppressor applications. The large Mylar 0.56µf at 400VDC capacitor, visible in Figure 4-9, is part of the over/undervoltage monitoring circuit.

Relay

The Song Chuan 832A-1A-F-S 48VDC relay is the most prominent component on the input side of the board, as shown in Figure 4-9. The single-pole, normally open relay has silver-tin contacts rated at 250VAC and 30A. According to the Song Chaun relay catalog (www.songchuan.de/en), the 40V coil requires 18mA for activation—that's about 1W to keep the relay energized. Of note are the maximum contact closure time of 15 millisecond (ms) and the 10ms maximum release time. The relay is marketed as a general-purpose power relay for power supplies and home appliances.

Circuit Board

The spacious, sturdy, double-sided circuit board is well laid out, with plenty of room for heat dissipation. My only criticism of the board is related to the placement of the holes for securing the large inductor. While the trace for the hot lead is more than an inch wide in places, it comes to a narrow of about 0.08 inch (2mm) because of a mounting hole drilled through the trace. A nonconductive center bolt that leaves the trace intact is probably a better mounting option.

How It Works

Using the simplified schematic in Figure 4-11 as a guide, let's focus on the circuits related to surge protection. The 12V dimmer circuit for the LED lighting is

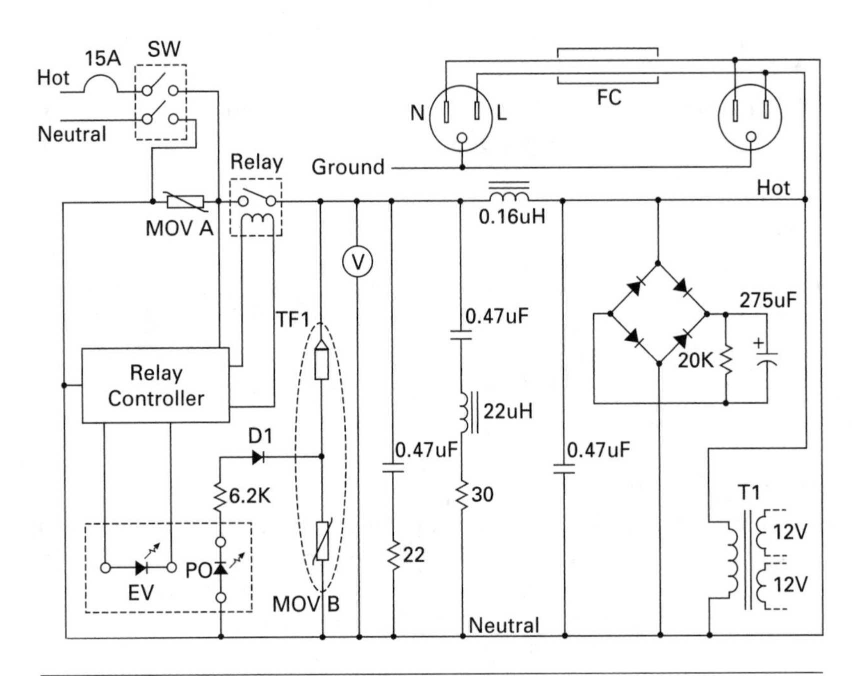

FIGURE 4-11 Simplified schematic

represented by the connection for the primary of transformer T1, between the line and neutral, at the bottom-right corner of the schematic. Similarly, the front-panel LED voltmeter is signified by the encircled *V* between hot and ground.

Power enters from the top-left and power is output through the two NEMA 15 outlets on the top-right. Each jack represents one "bank," as described on the PL-Plus package. The tubular ferrite core (FC) that encircles the line and neutral wires effectively adds inductance to the wire and "isolates" the three final sockets in the chain from the other six sockets. The amount of isolation provided by the ferrite core is not specified.

The hot lead is connected to a 15A circuit breaker prior to the DPDT on–off switch (SW) that interrupts both hot and neutral. After the on–off switch, directly across the hot and neutral lines is MOV A, the 20D391K-N, which has a maximum clamping voltage of 650V. If the MOV shorts, the on–off switch will be subject to short current until the 15A circuit breaker trips. With a maximum clamping voltage of 650V, the MOV is more likely to survive repeated voltage spikes than, say, a MOV rated at 300V, but it will also allow more spike energy to pass unabated.

The relay controller activates the normally open relay when the hot-to-neutral voltage is between about 90 and 140VAC. At voltages in excess of 140VAC, the relay releases, and the red "Extreme Voltage" LED (EV) is illuminated on the front panel. Because of the response time of the relay—a maximum of 10ms to release the contacts—and the relay control circuitry, the relay controller circuit doesn't come into play with subnanosecond voltage spikes.

The brunt of any spike that makes it past the relay is handled by MOV B, which has both a series thermal fuse (TF1) rated at 115°C and a monitoring LED (PO). The silicon power diode (D1) and 6.2KΩ resistor provide current to the green "Protection On" LED (PO) on the front panel. If the MOV overheats and TF1 opens, the LED will not illuminate. In this way, "Protection On" reflects the status of MOV B.

Moving to the right in the schematic is the LED voltmeter (V), which draws only a few milliamperes. The next component across the hot and neutral lines is a 0.47µf capacitor in series with a 22Ω, 1W fusible resistor. Given the capacitive reactance of the capacitor at 60Hz—about 6KΩ—the contribution of the 22Ω resistor is negligible. It's apparently in the circuit as a fuse, in the event the capacitor shorts secondary to damage from a transient. Recall that the MOV A clamps at 650V, which is significantly above the 275VAC rating of the capacitors.

The next circuit, just prior to the large choke in the line wire, is a series circuit composed of a 0.47µf capacitor, a 22µH inductor, and a 30Ω, 1W fusible resistor. Again the resistor apparently serves as a fuse. The inclusion of a small inductor, which, according to my Mouser Electronics catalog, has a maximum current rating of about 100mA, is curious at best. The resonant frequency of the 0.47µf capacitor and 22µH inductor is about 50kHz. That is, 50kHz signals between the hot and neutral wires are shunted, while signals greater or less than 50kHz pass and are not attenuated. This apparent tuning might provide some small improvement in attenuation of 50kHz signals—one of the test frequencies listed in the noise suppression test figures—but at the cost of increasing noise for all other frequencies, relative to a bare capacitor.

As noted earlier, the series inductor, the centerpiece of the power circuit board, provides 0.16µH of inductance, based on my measurements. Depending on the nature of the voltage spike, the inductor can both reduce the amplitude and lengthen the duration of a spike. This may be the rationale for including a 0.47µf capacitor directly across the line and neutral wires after the inductor.

The final circuit element is a diode bridge that feeds a 275µf at 250VDC electrolytic capacitor that leads to a 20KΩ bleeder resistor. This circuit, which I assume represents the manufacturer's linear filtering technology, is in my opinion of little value in protecting sensitive equipment plugged into the PL-Plus from significant voltage spikes.

While the diodes are capable of significant peak forward current from a spike, the current handling capabilities of the capacitor are limited. One of several reasons is that the function of relatively large value electrolytic capacitors degrades with increasing frequency, due to self-inductance. That's why electrolytic capacitors are often bypassed with relatively small value ceramic disc or Mylar capacitors.

As a source of protection from voltage spikes, the PL-Plus is essentially a pair of MOVs across the hot and neutral lines. The large inductor and, to a much lesser extent, linear power supply components probably absorb some of the energy of a transient. The question is how much energy is absorbed. Unlike the AV600, the "Protection On" indicator actually indicates the status of at least one of the MOVs. However, there is no protection from common-mode spikes or spikes from neutral to ground.

Mods

As with the AV600, I'm hesitant to recommend mods to line-powered equipment that's going to be in constant service. However, in my opinion, the PL-Plus is a platform to build on. Following are a few ideas for you to consider.

TMOV Upgrade

The fastest and easiest mod is to upgrade the first, exposed MOV to a fused MOV with a significantly lower clamping voltage rating. I picked up a Littelfuse 20mm TMOV (thermal MOV) with built-in thermal fuse, TMOV20R115E, from Mouser (www.mouser.com) for $1.30. The TMOV is rated at 115VAC continuous, with a clamping voltage of 300V, a 52J energy capacity, and a peak current of 10,000A. Littelfuse also offers a three-lead version, with a lead between the internal MOV and thermal resistor, for externally monitoring the state of the fuse. Both are compliant with UL 1449 2nd Edition.

Add Neutral and Ground MOVs

Install a small circuit board with Littelfuse 20mm TMOVs between ground and hot and ground and neutral. Whether or not "ground contamination" is a real problem for your equipment depends on whether you use equipment with switching power supplies or linear supplies, among other factors.

Component Deletions

Consider replacing the 22μH inductor with a wire jumper. I can see no reason for a filter limited to 50kHz.

Comparison

I don't want to turn this teardown into a product review, but I think it's useful to compare what we've learned about the AV600 and PL-Plus, realizing that they're very different products serving different markets. Based on the teardowns, my overall assessment is that while the PL-Plus is certainly a more robust device, at home with the coolest looking AV equipment on the market, the little plastic AV600 is more sensitive and offers more spike protection per dollar.

The PL-Plus is obviously superior to the AV600 as a power distribution unit and in the areas of over- and undervoltage protection and noise suppression. Overvoltage could be a problem if you have an emergency generator with an improper output setting, or if you generate your own power from solar cells or a wind turbine. In addition, if you have a band that frequently sets up around fluorescent lights, the noise suppression offered by the PL-Plus could be invaluable.

While the PL-Plus clearly has a place next to high-end AV equipment (I now own three Furman power conditioners), the findings of the teardown don't support—and in some cases refute—many of the claims made by Furman. At best, many of the statements in the marketing materials are misleading. For example, the components we discovered during the teardown simply don't support the claim of a spike-clamping voltage of 133VAC RMS. The claim that the circuitry includes both a 12,000A MOV and a 12,000A TVZ was clearly refuted. In addition, the relative effectiveness of the linear filtering circuitry is unsubstantiated.

The specifications of the AV600 are also suspect. For example, I was unable to locate the MOV datasheet to verify the 1ns response time. That the "Protection On" LED functions with the ground wire disconnected and every MOV in the unit blown is problematic.

One issue I have with the PL-Plus as a user is the lack of UL 1449 compliance. Among other requirements, the latest revision requires specific safety measures, such as thermally fused protection of MOVs. It's not clear if the AV600 will pass the latest UL 1449 requirements without further testing or modification, but it's clearly designed to disengage the MOVs quickly if they overheat.

Chapter 5

Electronic Pedometer

Home entertainment systems have been cited as one reason for the expanding American waistline. That may be, but electronic devices are also helping us trim the fat. A simple but effective electronic aid to fitness that you may own or have access to is the electronic pedometer. In its simplest expression, a pedometer is an instrument that counts the number of steps you take. Advanced units calculate calories burned as they track your pace and time. Try counting your steps as you walk or run and you'll immediately appreciate the utility of a pedometer.

In this chapter, I'll tear down an Omron HJ-112 electronic pedometer, shown in Figure 5-1. This digital meter, which retails for about $40, is widely considered one of the top instruments on the market. In addition to discussing to technologies used in competing products, I'll suggest an experimental mod that can extend the basic functionality of the unit.

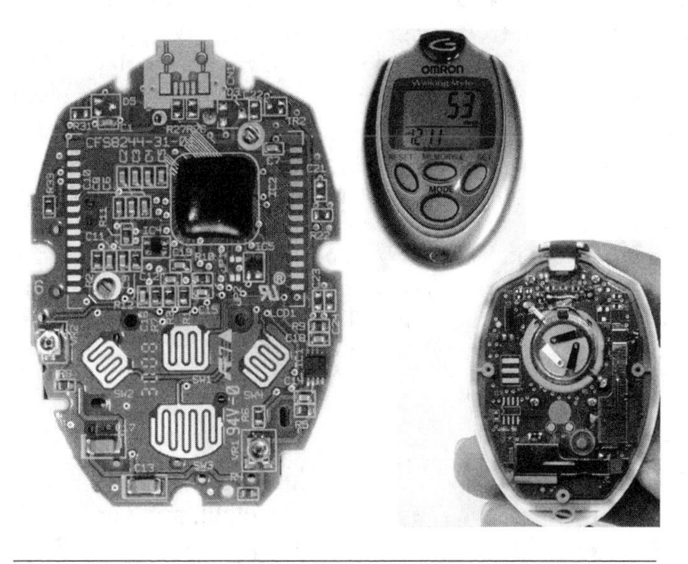

FIGURE 5-1 Omron HJ-112 electronic pedometer

Highlights

I was initially disappointed with this teardown. After reading the marketing copy and rave reviews for the Omron HJ-112, hereafter referred to as the Omron, I expected an all-electronic pedometer, complete with the latest solid-state micro electromechanical systems (MEMS) accelerometers. However, when I first cracked the case and found what appeared to be old-fashioned pendulum switches, I thought of abandoning the teardown. I changed my mind a few minutes later after examining the circuitry and function of the sensors.

This teardown explores an integrated electromechanical system that combines SMT components, a microcontroller that I'll largely ignore, three peripheral chips, and a pair of unique electromechanical sensors. During the teardown, note the following:

- The use of SMT components, including resistors, capacitors, and integrated circuits (ICs)
- The LCD panel, including the mounting hardware
- The tuning fork crystal oscillator
- The construction of the pendulum sensors
- The silicone-carbon switch components

Specifications

The 2.8 × 2.1 × 0.6 inch (HWD) pedometer measures steps with an accuracy of ±5 percent. That is, if you take 1000 steps, the count can be as high as 1050 or as low as 950 and still be within specifications. Based on this step count and elapsed time, the microcontroller calculates and displays calories, distance, and pace. It also displays aerobic steps and minutes, which are the number of steps and elapsed time since 10 minutes of continuous activity. Power is provide by a single CR2032 3V lithium cell, which is rated at six months, assuming 10,000 steps per day. The 1 ounce, Chinese-made Omron has a seven-day memory.

Because the calculation of calories is based on the step count, the calorie counter function is probably no more than ±5 percent accurate. In addition, because age is not used to calculate calories burned, you have to multiply the displayed figure by an age-dependent correction factor from a table in the user manual for an accurate estimate of caloric expenditure. The default equation assumes either a man in his 60s or a woman in her 30s.

Operation

Setup involves entering the time, your weight, and your stride length using the four elastomeric buttons. After that, it's simply a matter of clipping the unit to your belt, backpack, or purse. However, the unit must be perpendicular to the ground during operation. In other words, you can't simply toss the unit into your backpack or briefcase. This is a major limitation of using a dual-axis sensor system.

Teardown

If you're following along with a unit at home, you should be able to complete the basic teardown, illustrated in Figure 5-2, in less than 5 minutes.

Tools and Instruments

You'll need a miniature screwdriver to crack the case and a good magnifier to examine the SMT components. Have a soldering iron handy if you want to examine the two pendulum sensors in detail.

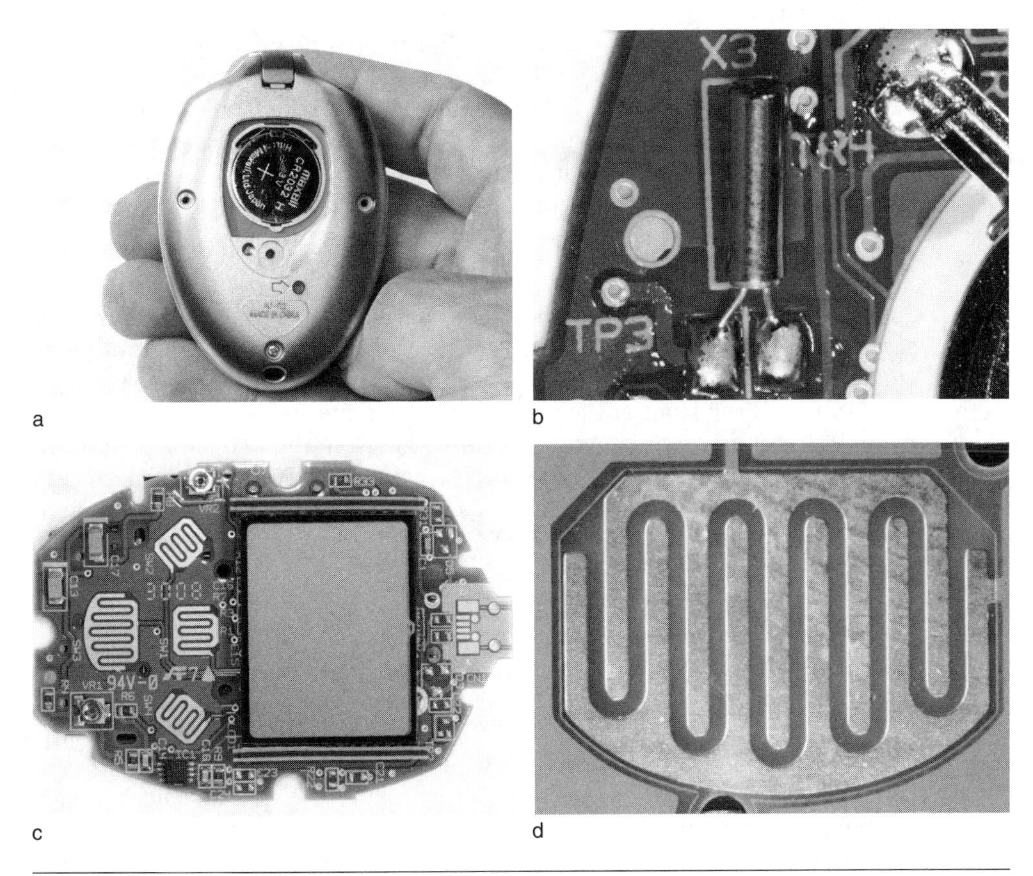

FIGURE 5-2 Teardown sequence

Step by Step

If you plan to use the pedometer after the teardown, you should probably stop after viewing the back of the circuit board. Once you remove the circuit board completely from the plastic shell, your odds of reseating the LCD panel are slim.

Step 1

Remove the 3V lithium cell. Use a miniature Phillips-head screwdriver to remove the single screw securing the small battery hatch, as shown in Figure 5-2a.

Step 2

Remove the back. Remove the three Phillips-head screws securing the back cover, shown in Figure 5-2a. With one hand, pull up and out on the plastic retention hook at the top of the unit and pull up on the back cover to reveal the back of the circuit board, shown in the lower-right corner of Figure 5-1.

Note the thin rubber gasket that runs in the groove along the periphery of the case. If you plan to use the pedometer after the teardown, do your best to avoid damaging the gasket. Also note the two metal covers of the pendulum switches, immediately to the right and below the lithium cell contacts. Note the silicone reset switch between the two mechanical pendulum switches, and the cylindrical tuning fork crystal oscillator, at the 10 o'clock position relative to the cell contacts. A close-up of the crystal oscillator is shown in Figure 5-2b.

Step 3

Separate the circuit board from the front of the case. Remove the four Phillips-head screws above and below the cell contacts, visible in the lower-right of Figure 5-1. Hold the board in place while you flip the unit over. Gently lift the plastic front plate to reveal the front of the circuit board. Rotate the board so that the LCD panel is on the right, as shown in Figure 5-2c. Note the two SMT potentiometers at the 8 o'clock and 11 o'clock positions relative to the center of the board. Also note the four serpentine pads for silicone-carbon elastomeric switches, which occupy much of the board to the left of the LCD panel. A close-up of a pad is shown in Figure 5-2d.

Now examine how the LCD panel is attached to the board. Referencing Figure 5-2c, note the gray Zebra elastomeric connector that runs along the sides of the LCD panel. As discussed in detail in the teardown of the digital bathroom scale (Chapter 3), the Zebra connector provides hundreds of parallel, electrically insulated connections between the LCD panel and the foil pads on the circuit board directly beneath the panel. One of the many advantages of the Zebra connector over alternative connection technologies is that no solder or special spring mount system is required to connect an LCD to a circuit board. The downside is that reassembling the panel back without a jig to hold everything in place is a challenge.

Step 4

Remove the LCD panel. Secure the circuit board and carefully lift up on the LCD panel. The Zebra connectors may adhere to the circuit board or the LCD panel, or they may simply fall off as you lift the panel. Place the connectors and the panel in a clean, dust-free container. Note the additional components visible, including three ICs and several SMT resistors and capacitors, as shown in Figure 5-1. A microcontroller is hidden under the black epoxy blob.

Layout

As revealed in the teardown, the layout is simple, clean, and compact. Solder joints are well executed on even the smallest SMT components.

The challenge presented by this board is that the components and traces are so small that you need a good magnifier and a multimeter with a set of sharp probes to explore the circuitry. Moreover, many of the traces are hidden under an insulating layer between the battery and circuit board. Fortunately, there are multiple test points on the circuit board to facilitate testing and circuit tracing.

Components

Of the components in this teardown, the most intriguing are the pendulum sensors. The only other consumer devices that come to mind that might include similar switches are car alarms and motion detector alarms for laptops.

Pendulum Sensors

The two pendulum sensors, shown in Figures 5-1, 5-3a, and 5-3b, are aptly named because each sensor consists of a flexible shaft attached to the circuit board at one end and to a broad copper wing at the other end. As shown in Figure 5-3a, the wing-shaped ends of each pendulum are confined to a metal enclosure. If you shake the board in the plane of the pendulum, the copper wing-shaped ends move slightly with no perceptible oscillation. The motion of your torso as you walk should result in the same motion.

Figure 5-3b shows one of the pendulum sensors without its cover. The foam rubber glued to either side of the pendulum arm dampens the mechanical oscillation of the pendulum. Without the foam and the wing-shaped air dampers, each pendulum could move for several seconds and falsely elevate step count.

At first I thought that the copper wings at the end of each pendulum sensor made contact with the grounded enclosure. Then I noticed that the solder connection from the circuit board to the arms of each pendulum isn't a direct connection. As shown in Figure 5-4, a short wire is soldered to what appears to be a length of double-silvered mica. As discussed next, it took a bit of investigation to discover what

FIGURE 5-3 Pendulum sensor

FIGURE 5-4 Pendulum sensor connection

was really going on with the pendulum sensors. If you're working on a pedometer at home, stop reading and try to figure out how the sensors work on your own.

Microcontroller

The microcontroller hidden under the glob of black epoxy, visible in Figure 5-1, handles the LCD display and the momentary switches, and it reads signals from the sensors. Review the discussion of how to remove epoxy blobs in Chapter 3, and you'll see that if you remove the epoxy in this case, you may be able to visualize the raw silicon chip, but that's about it. For our purposes, what's important is that the microcontroller reads the two pendulum sensors, processes the data, and displays the results.

Microcontroller Support ICs

In addition to an external crystal oscillator, the microcontroller relies on three external ICs for normal operation—two CMOS (complementary metal oxide semiconductor) voltage detectors and a dual operational amplifier. The B8K and CUN voltage detectors are packaged as 4-pin SC-82AB devices, as shown in Figure 5-5a. It's difficult to see, but pin 2, in the upper-left corner of the device, is slightly wider than the other three pins. The 8-pin dual operational amplifier is in a standard SOP-8 package, shown in Figure 5-5b.

The voltage detectors, which connect directly to the microcontroller, are triggered when the battery voltage drops to a preset level. The operational amplifiers are used to condition the signals from the pendulum sensors.

Crystal Oscillator

The cylindrical tuning fork crystal oscillator, which operates at 32.768kHz, connects directly to the microcontroller. Tuning fork crystal oscillators are popular in portable devices that need a real-time clock because the oscillator package is small, the operating frequency is stable over a relatively wide temperature range, and the fundamental frequency is low. Clock frequency is a compromise between performance and battery life.

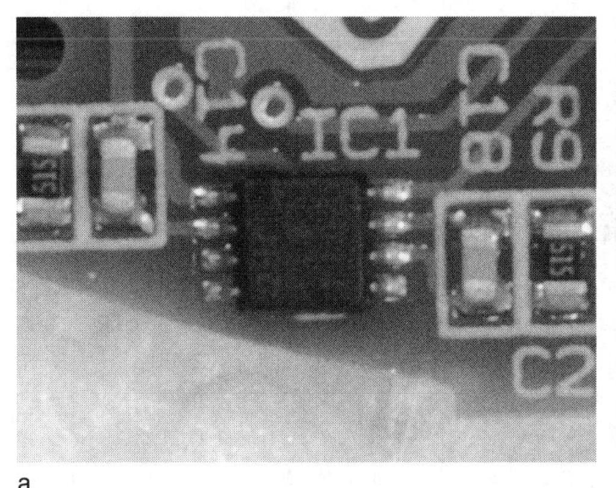

a

b

FIGURE 5-5 Microcontroller support ICs

SMT Potentiometers

The two 14KΩ linear potentiometers, shown as VR1 and VR2 in the lower half of the large circuit board shone in Figure 5-1, are used to set the sensitivity of the pendulum sensors. These SMT potentiometers are unlike typical full-sized potentiometers in that the wiper arm is free to rotate continuously. If you rotate the Phillips-head screw clockwise, the resistance from the wiper arm to one of the end terminals cycles from 0 through 14KΩ. The lack of a mechanical stop is common in SMT potentiometers.

LCD Panel

The LCD panel is notable in its use of dual Zebra elastomeric connectors to mate it with the circuit board. The connectors not only facilitate automated assembly of the device, but they protect the panel from vibration and g-forces. Given that the device is likely to be dropped, this protection increases the odds of the panel surviving an impact with the ground.

Silicone-Carbon Elastomeric Buttons

The pedometer relies on four silicone-carbon buttons: a reset button on the back of the circuit board, and buttons for mode, history, set, and reset functions on the front of the pedometer. The bottom of each button is covered with conductive carbon that, when pressed, shorts an underlying circuit board trace, such as the one shown in Figure 5-2d, completing the circuit.

Elastomeric buttons are lightweight and inexpensive, and they provide excellent tactile feedback in a small package. In addition, the design of the silicone button assembly adds to the water-resistance of the pedometer. A related switch design that uses circuit board tracings as part of the circuit is discussed in Chapter 10.

How It Works

The major components and their contributions to the pedometer are shown in the simplified schematic of Figure 5-6. Power is supplied by a single 3V lithium cell that is connected directly to the microcontroller and supporting ICs. Depressing one of the five elastomeric switches, represented by the single switch (S1–5), activates the microcontroller and the LCD. The crystal oscillator (XT) provides the microcontroller with a stable clock signal, which is required for the clock function and to calculate pace.

The horizontal (TH) and vertical (TV) piezoelectric pendulum transducers generate an irregular, alternating signal, as a function of the wearer's motion. IC1 acts as a signal conditioner, producing a clean pulse that's easily processed by the analog-to-digital conversion circuitry within the microcontroller. The 14KΩ SMT

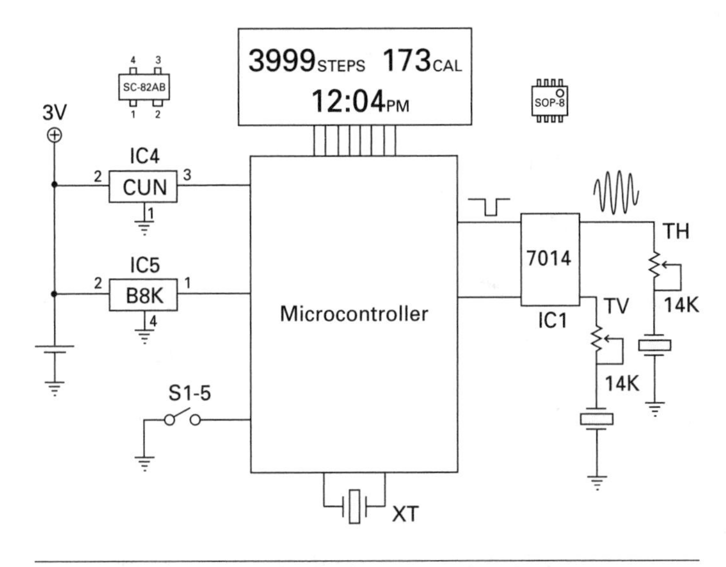

FIGURE 5-6 Simplified schematic of pedometer

potentiometers in series with the piezoelectric transducers are used to control for differences in transducer output. This form of signal conditioning should be familiar to you by now, since I covered the same basic circuit in the teardown of the digital bathroom scale in Chapter 3.

To appreciate what the pulse shaping circuit has to handle, see Figure 5-7, which shows the signal from the horizontal pendulum transducer during one of my steps with the Omron clipped to my belt. As you can see, a single step results in a 500mV signal with duration of about 100ms.

One of the challenges in the design of battery-operated, computerized devices is how to handle a nearly depleted battery at startup and in the middle of an operation. Think of your laptop: it warns you of impending battery death and initiates an automatic shutdown sequence. It also refuses to boot up if the battery is depleted.

As shown in the schematic, the two voltage detectors connect directly to the microcontroller and provide it with information on the state of the battery. IC5, the B8K voltage detector, is triggered at 2.5VDC. IC4, the CUN detector, triggers at about 2.0VDC. Without access to the source code, it's difficult to determine exactly what the microcontroller does with the information provided by the voltage detectors. My assumption is that IC5 triggers a controlled shutdown procedure that protects the data and the state of the microcontroller registers. The state of the CUN voltage detector is probably checked early in the microcontroller startup sequence. If the battery voltage is less than 2.0V, the startup sequence is aborted, thereby avoiding a crash during startup, which could corrupt the registers.

An alternative to the CMOS voltage detectors is to use one or more analog-to-digital converters within the microcontroller to monitor battery voltage periodically. The downside of this solution is the significant overhead of an additional analog channel, in terms of chip complexity, cost, and power requirements. Not only do the

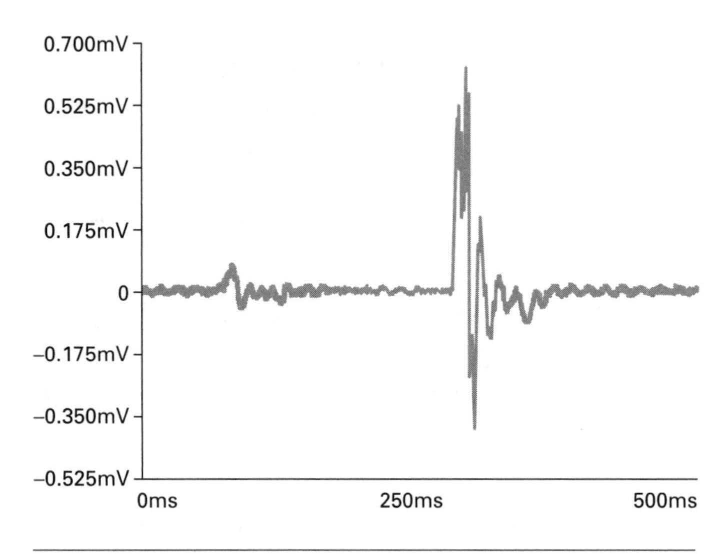

FIGURE 5-7 Raw pendulum sensor output

external voltage detectors provide superior accuracy, but the current requirements are only 0.8µA.

Alternative Technologies

Based on my use, the piezoelectric pendulum transducers are more than sufficient for the pedometer. However, there are alternative sensor technologies that range from cheaper, simpler motion sensors, to more expensive and more advanced transducers that can be integrated with other useful functionality.

Single-axis Pendulum

A simple, inexpensive alternative to the dual-axis piezoelectric pendulum is a single-axis mechanical pendulum, such as the one used in the Korean-made MS-2000 digital counter (see Figure 5-8, left). If you examine the image, you will see the mechanical pendulum switch, including the single wire spring that can be set at one of three notches to adjust for stride length.

Note that in normal use, the pedometer should be worn so that the pendulum arm is horizontal to the ground. When the wearer walks, the weighted pendulum bobs up and down, and in so doing, it makes mechanical contact with the second electrode. Note the black epoxy blob that hides the microprocessor that counts the steps and displays the total count. The unit has a tiny, four-digit LCD display and a single button to zero the count. There is no calculation of calories or an electronic method of specifying weight or stride. The only user-customizable setting is the spring tension.

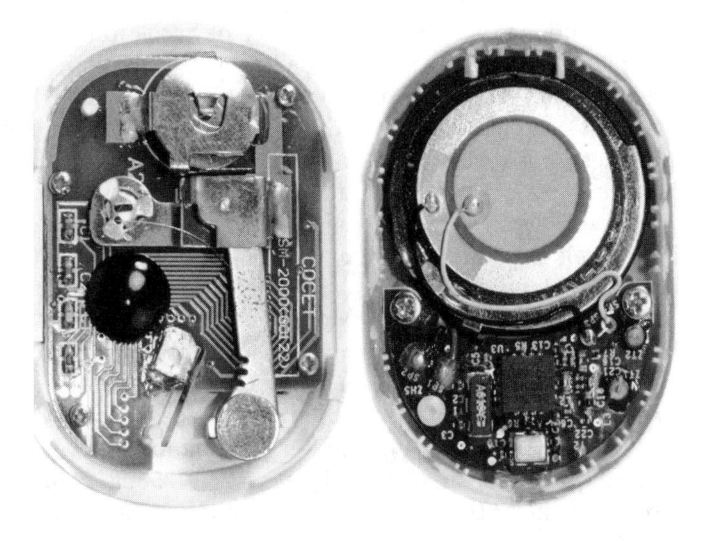

FIGURE 5-8 Mechanical pendulum (left) and piezoelectric step sensor (right)

Piezoelectric Sound Detector

Related to our piezoelectric pendulum sensors is the piezoelectric sound sensor used in the Apple/Nike shoe sensor, shown on the right in Figure 5-8. (Note that the two devices in the image are not to scale. The Apple/Nike sensor is about half the size of the MS-2000 pedometer.) The piezoelectric sensor, which sends signals wirelessly to an iPod with a receiver, triggers on the sound produced by the wearer's foot slapping the ground. Nike sells shoes with built-in holders in the soles. However, I've used the sensor attached to other shoes with good results.

I suspect the sensor would have problems if the wearer runs on loose gravel or soft grass. One of the common complaints of the Nike/Apple sensor is that the case is sealed and that there is no way to change the battery. I used diagonal cutters to nibble away the red and white top to get a view of the sensor and wireless communications circuitry. The sensor should remind you of the bathroom scale teardown in Chapter 3, in which we discovered a piezoelectric sensor used to turn on the scale.

MEMS Accelerometer

If you have an Apple iPhone/iTouch, a Sony/Ericsson w710i phone, or one of several other cell phones, you have at least the potential of owning a pedometer. These devices include two or more multiaxis MEMS accelerometers that can be accessed by software, transforming the device into an expensive but readily available pedometer.

From a functional perspective, the two-axis accelerometer in the Sony/Ericsson phone offers little advantage over our pedometer, other than convenience. The real advantage of MEMS accelerometers is obvious with the Apple iPhone, which sports a three-axis MEMS accelerometer. With three axes covered, the iPhone can respond to steps regardless of its orientation. It can be upside down in a backpack, flat, or at any angle relative to the ground. The limitation of the iPhone, of course, is the initial cost of the phone and the need to buy and install a pedometer app.

MEMS accelerometers vary in mode of operation, but one that I've worked with extensively, the Memsic 2125, uses a heater to warm a bubble of gas that moves in response to gravity and acceleration of the device. Thermopiles near the periphery of the bubble sense identical temperatures when the accelerometer is level and different temperatures when the sensor is tilted. Temperature differences detected by the thermopiles are converted into pulses by the onboard electronics and the angle of tilt is calculated using elementary trigonometry.

If you want to experiment with the Memsic 2125, it's available from www.parallax.com for about the price of our pedometer. And this is the rub. In the current pedometer market, the cost of a three-axis MEMS and peripheral circuitry is prohibitive.

Mods

Given the compact, integrated design, there really aren't many mods that come to mind that could significantly improve upon this pedometer. However, if you're willing to give up portability, the pedometer makes a great platform to explore multiaxis signal detection. If all else fails, the pedometer is a good source for SMT components, and for the piezoelectric pendulum sensors.

Add a Third Axis

A simple bench experiment is to add a third piezoelectric sensor in parallel with one of the existing pendulum sensors. The challenge is presenting the unit with a signal source with amplitude and spectral characteristics compatible with the piezoelectric pendulum sensors. You could try to roll your own third axis sensor with a piezo speaker element, or harvest the sensor from a second Omron pedometer.

Chapter 6

Compact Fluorescent Lamp

One of the casualties of the drive toward carbon neutrality is the tungsten light bulb. LED, halogen, and fluorescent lights are increasingly popular alternatives to the inefficient incandescent bulbs, in part because local governments and utility companies subsidize these green technologies, and in part because they're theoretically cheaper to operate and have a significantly longer lifespan. A rate-limiting step in the move to higher efficiency lighting is the infrastructure of existing screw-in sockets.

The form and function limitations imposed by screw-in sockets have been partially addressed by the compact fluorescent (CF) lamp, such as the one shown partially disassembled in Figure 6-1. To fit and operate in regular sockets and fixtures, the base of CF lamps designed for residential use contains disposable

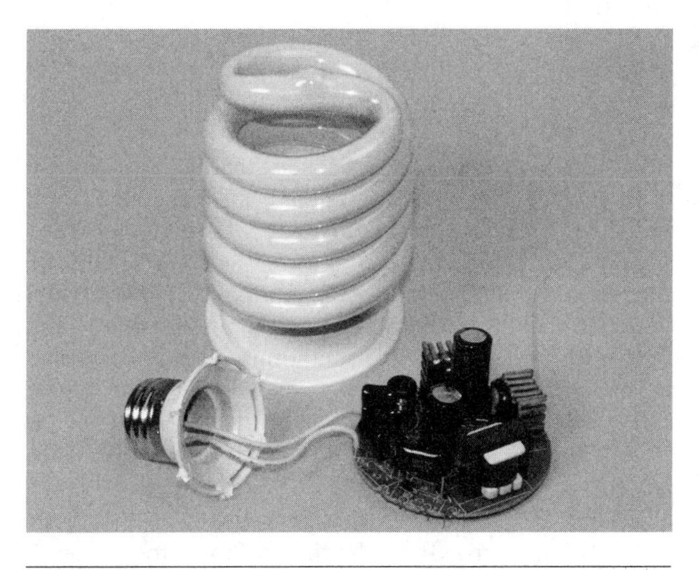

FIGURE 6-1 Compact fluorescent lamp, partially disassembled

ballast circuitry. As such, CF lamps are a bridging technology that will do for now. CF lamps are not as efficient or cost effective as the regular straight fluorescent lamps used in offices and factories. In addition, the production, assembly, and disposal of the electronic components add to the carbon footprint of CF lamps. Disposing of CF lamps also increases the load of mercury, lead, arsenic, and other toxins on the environment.

In this chapter, we'll tear down the 42W TCP SpringLamp shown in Figure 6-1. This CF lamp is available online from vendors such as bulbs.com for less than $20. Although you may not have this particular lamp at home, the layout and component functions differ surprisingly little from one brand to another.

Highlights

If you've never cracked open a CF bulb, you're in for a pleasant surprise. There's quite a bit of technology packed in the base of each bulb. I've opened dozens of different models, and every one had two switching transistors, a toroidal transformer, several inductors, several diodes, a handful of capacitors, and a few resistors. As I noted earlier, it seems paradoxical to create a disposable technology in the name of saving the planet. However, if you're into repurposing hardware, you're the immediate beneficiary of this low-cost source of components.

Now for the disclaimer. It's a good idea to think of the spiral bulb as a snake coiled to strike. If you're careless and unlucky, you could get cut, poisoned, and electrocuted—all within a few seconds. So it's important to keep the bulb intact.

Whether you follow along virtually or with your own CF, during the teardown, note the following:

- **The component types** Do you see any SMT components? If not, why not, given the cramped conditions in the base of the bulb?
- **Provisions for component failure** Are there any fuses or components designed to fail gracefully?
- **The thermal design** Most CF lamps can't operate in closed environments because of heat buildup, and heat kills components. Moreover, many CF lamps can operate only in certain orientations, such as base-up. Are there any heat sinks? Vents? Any active cooling?

Specifications

A typical 100W soft white incandescent bulb is rated at 1500 lumens and 1200 hours. The "soft white" designation means that the color temperature of the lamp is 2700°K. *Color temperature* refers to the color of the light emitted by the

Dangers: Mercury Toxicity and High Voltage

If you apply a little too much pressure to the bulb, it will break. If you're not wearing gloves and the sharp end of the bulb pierces the palm of your hand, it's probably off to the ER for stitches. Leather gloves and eye protection may protect you from glass shards, but they won't protect you from the danger of mercury poisoning. The bulb is filled with a gas mixture composed of mercury vapor, argon, xenon, and neon at a pressure of about 0.003 atmospheres.

To give you an idea of the toxicity of the mercury contained in a CF bulb, consider the recommendation of the US Environmental Protection Agency (EPA). According to the EPA, if you break a CF bulb, you should open a window and leave the room for at least 15 minutes. You should also shut off the central heating or air conditioner to avoid the spread of mercury vapors. When you return, use duct tape to pick up the shards of glass—don't launch the shards into the air with a vacuum cleaner. *Throw away* any clothing that has been in direct contact with the shards. Washing the clothes in machine washer will *contaminate the machine*.

In case you're curious, the effects of mercury toxicity typically aren't immediately apparent after exposure. Symptoms usually appear years after chronic exposure, but a single large exposure can cause immediate problems. According to the National Institutes of Health, you'll likely develop significant neurological deficits—inability to walk, blindness, memory problems, uncontrollable shake, and seizures. Not something you want for yourself or your family.

The other potential health hazard, high voltage, has an immediate effect. If you decide to trace the live circuitry, you're minimally exposing yourself to 120VAC. If you work with a powerful CF lamp, such as the 42W lamp featured in this teardown, you'll be exposed to 600VAC or more. You should have high-voltage equipment and tools on hand if you're working with a live CF bulb.

The bottom line is that if you don't have the protective clothing, tools, and experience, then either enjoy the teardown vicariously or find someone with the skills and tools to work with you.

bulb. The 2700°K is equivalent to the yellowish light produced by a tungsten bulb. 1500 lumens is the amount of light output. Average lifetime is the average service lifetime of a 100W bulb. It's not the average lifetime you can expect from your particular bulb.

In contrast, the specifications of the TCP 42W SpringLamp are as follows:

- **Color temperature:** 5100°K/bright white
- **Average lifetime:** 10,000 hours
- **Lumens:** 2800
- **Base:** Medium screw (E26)
- **Input line voltage:** 120VAC
- **Minimum starting temperature:** –20°F
- **Energy used:** 42W

- **Incandescent equivalent:** 150W
- **Power factor:** >50 percent
- **CRI (color rendering index, aka color temperature):** 84
- **THD (total harmonic distortion):** <150 percent
- **Length:** 7.0 inches
- **Diameter:** 2.8 inches

Although most of these terms should be familiar to you, this list of specifications hints at the complexity of the typical CF relative to the simple tungsten bulb. The SpringLamp features a 5100°K bulb and emits relatively white light. A related specification is the CRI, which relates to how well the light renders colors naturally. A CRI of 100 is perfect. The SpringLamp, like most CF lamps, doesn't produce much red light, so reds are muddied. As a result, the CRI score is only 84—good but not great.

Unlike tungsten lamps, CF lamps employ an initial starting arc, generally assisted with heating elements in the ends of the bulb. The indoor SpringLamp won't start below –20°F—meaning you probably wouldn't want to install it in a walk-in freezer.

The lumens and incandescent equivalent specifications relate the CF to a standard tungsten bulb. For about the cost of running a 42W incandescent bulb, you get 150W of light—or more specifically, 2800 lumens. I qualify the cost because of the horrible power factor figure of >50 percent.

An incandescent bulb, which is simply a resistor that releases about 10 percent of the dissipated energy as light, has a power factor of essentially 100 percent. That is, the real power—in watts—is the same as the apparent power—volts × amps. In contrast, this CF has a power factor of >50 percent, which means that the volt-amp figure can be 50 percent greater than suggested by the 42W specification. The relationship between real power (watts), apparent power (volts × amps), and power factor can be expressed mathematically as follows:

$$\text{Power factor} = (\text{input watts}) / (\text{volts} \times \text{amps})$$

Given our 42W CF and a line voltage of 120VAC, what can we expect for the actual input current? If we were dealing with a resistor or incandescent bulb, each of which has a power factor of 1.0 or 100 percent, the current would be as follows:

$$\text{Input watts} = \text{power factor} \times \text{volts} \times \text{amps}$$

$$42\text{W} = 1 \times 120\text{V} \times \text{amps}$$

$$\text{Amps} = 42\text{W}/120\text{V}$$

$$\text{Amps} = 0.35$$

So, with a purely resistive load, and, by definition, a power factor of 100 percent, a 42W bulb would draw 0.35A. Now, with our CF lamp and a power factor of >50 percent (let's call it 50 percent for simplicity), we have this:

$$\text{Input watts} = \text{power factor} \times \text{volts} \times \text{amps}$$

$$42W = 0.5 \times 120V \times \text{amps}$$

$$\text{Amps} = 42W/(0.5 \times 120V)$$

$$\text{Amps} = 0.70$$

Therefore, the power line has to supply our CF bulb with 0.7A. This shouldn't be a major problem with only one bulb, but with a dozen bulbs in a house, it could be significant. The higher current increases power losses in the distribution systems—recall power lost in wires and components is calculated as $P = I^2R$—and can result in higher charges from the utility company. This charge isn't automatic, because most residential wattmeters measure real, not apparent, power. However, utility companies can easily measure the power factor and adjust your utility bill accordingly.

The reason that the CF has a power factor of only >50 percent is that the input circuit is nonlinear. As you'll see later, the input consists of a pair of diodes that convert 120VAC to DC. This nonlinearity results in the generation of harmonic currents that can wreak havoc on the power distribution system. Better CF lamps have a THD of less than 10 percent. The rating of less than 150 percent for this CF is disappointing. The THD and power factor specifications are interrelated: improving the power factor will improve the THD.

Operation

There is little to say about operating a CF lamp. When you supply 120VAC, the lamp preheats the gases in the bulb for a moment, and then the electronic ballast ignites the lamp with a high-voltage charge. The ballast maintains the proper current flow until the power is removed.

Most CF lamps require full line voltage at all times, so they can't be used with dimmer switches. Similarly, although there are exceptions, CF lamps require good ventilation and risk premature failure and fire if operated in sealed fixtures.

Another hazard when working around CF-generated light is exposure to ultraviolet (UV) radiation. When mercury vapor is ionized, it generates UV light, which excites phosphors in the wall of the spiral tube. These phosphors generate visible light. However, some of the UV light makes it through the tube unimpeded and causes damage to paint pigments, plastics, skin, and eyes. So, it's probably not a good idea to replace your halogen reading lamp with a CF lamp—even if you need to work on your tan.

Teardown

If you're good with your hands and the CF bulb that you're using isn't glued together, this is a 60-second teardown. Even if you have to resort to a hacksaw to open the plastic base, you shouldn't require more than 5 minutes start to finish. The real fun, of course, is figuring out how all those components work together to generate light.

Tools and Instruments

The basic teardown requires safety goggles, a pair of leather or heavy cotton work gloves to protect your hands from a broken fluorescent tube, and either strong fingernails or a standard screwdriver blade to pry apart the plastic housing. If you're working with a high wattage bulb (40W and above), you'll need a multimeter with a high-voltage probe to examine the live circuitry.

Most high-voltage probes are designed for a specific meter or series of meters. For example, I use a Fluke 80K-40HV probe with my Fluke 87 digital multimeter (DMM). The probe is designed so that when it's used with a meter with an input impedance of 10M, it functions as a 1000:1 divider. Figure 6-2 shows the simplified circuit diagram of the HV probe and a DMM with an input impedance of 10MΩ. The input impedance of my probe is about 79M, which is provided primarily by R2. R1 is about 80KΩ. R1 in parallel with the 10M input impedance works out to about 79KΩ, or a ratio of 79MΩ:79KΩ or 1000:1.

It's possible to use the Fluke or other HV probe with a multimeter with a lower input impedance than called for in the design. However, you'll have to multiply the displayed voltage by a correction factor because the displayed voltage will be lower than the actual voltage. See your probe's documentation to determine the correction factor for your multimeter.

The ground lead of the HV probe must be connected to earth ground for the divider to work properly. The closest earth ground is probably the ground pin of a power outlet. Earth ground isn't chassis ground, although a chassis could be

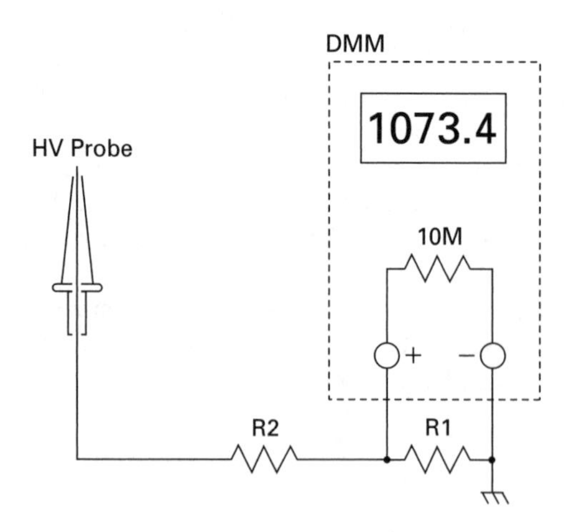

FIGURE 6-2 High-voltage probe circuit

connected to earth ground. If you fail to attach the ground lead of the HV probe to earth ground, you risk melting your DMM and possibly receiving a nasty shock if you serve as the ground return for the voltage you're trying to measure.

Step by Step

The teardown sequence shown in Figure 6-3 is more of a "crack open and look" than a real teardown. Even so, pay attention. If you happen to break a fluorescent tube, treat whatever is left of the tube as a biohazard and seal it in a plastic bag. Check with your public health office for how to best dispose of the hazardous waste.

Step 1

Open the base of the bulb. Grasp the top of the base—not the glass tube—with your gloved hand. Position your hand so that, should the glass bulb break, shards of glass would be blocked from your face and other hand. Now, with your other hand, use your nails, a standard screwdriver, a dime, or a similar thin tool to pry open the enclosure. I used my nails on the base, but you may want to use a screwdriver. Just don't run the screwdriver blade through the electronics and wires in the process of opening the base.

Another option is to use a hacksaw to cut away the plastic skin, right along the seam. However, remember that just below that thin, plastic skin are a dozen or so components. If you opt for the hacksaw approach, secure the lowest part of the base—the part farthest from the bulb—in a vise. And don't hit the glass tube with your hacksaw.

Regardless of the method you use, you should see something like Figure 6-3a when you first crack open the lamp.

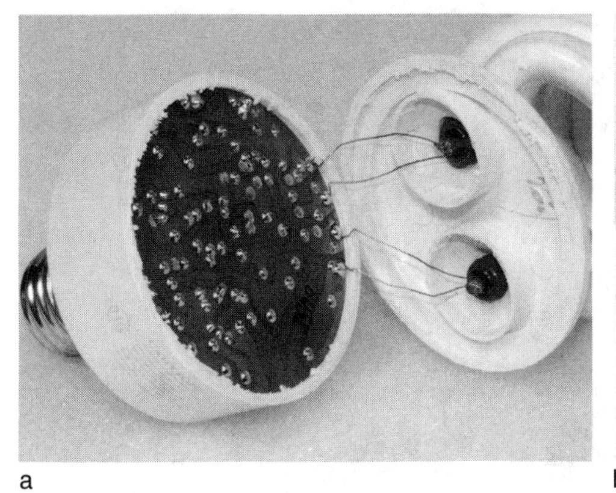

a

b

FIGURE 6-3 Teardown sequence

Step 2

Use a pair of diagonal cutters to sever the four wires between the bulb and the circuit board. Put the glass bulb aside for now. Examine the foil side of the circuit board, shown in Figure 6-3b. Note the generous spacing and thickness of the traces, as you'd expect for a high-voltage, high-current circuit.

Layout

The layout of every CF that I've cracked open is remarkably similar, despite differences in power and manufacturer. They all share the same basic layout as the populated circuit board shown in Figure 6-4.

As you can see, the circuit board is a disc, with the two switching transistors and inductors toward the periphery. The electrolytic capacitors are in the center of the disc. The silver mica capacitors, including the starter capacitor used to ignite the bulb, are located on the periphery of the board, along with wires to AC power and the bulb.

Components

Regardless of the make and model of your CF lamp, it probably has a spiral lamp with dual cathode filaments, a pair of switching transistors, a ferrite core toroidal transformer, a pair of inductors, a handful of silicon power diodes, a half-dozen capacitors, and a DIAC (diode for alternating current).

FIGURE 6-4 CF layout

Fluorescent Lamp

The spiral fluorescent lamp is filled with low-pressure mercury vapor that, when ionized by an electrical current, generates UV light. A mixture of phosphors in the tube absorbs this UV radiation and emits lower energy radiation in the visible spectrum.

Most CF bulbs designed for residential use have a rapid start system that requires a combination filament-cathode at each end of the spiral tube. The two 1.5Ω filaments in the SpringLamp CF are used to preheat the bulb for a few tenths of a second, lowering the voltage necessary for ionization of the low-pressure gas mixture.

Transistors and Heat Sinks

This CF features two D13009K NPN switching transistors rated at 400V collector-emitter voltage and 5A collector current. They're packaged in a standard TO-220 case, which has a metal tab extending up from the back of the transistor. As is standard for a TO-220 transistor, the leads are base, collector, emitter, looking from the front of the device. Figure 6-5 shows multifingered heat sinks bolted to the metal tab of each transistor. The heat sinks increase the heat dissipation from each transistor via natural convective cooling. Note the 100mH inductor on the left and a toroidal transformer on the right.

This is a good point to focus on the heat sinks used in this CF and on transistors in general. The goal of using a heat sink is to keep the transistor junction below the maximum rated temperature. But how much of what kind of heat sink is enough?

FIGURE 6-5 NPN switching transistor with heat sink

If you search the Web and basic electronics texts, you'll find dozens of formulas that can be used to define the optimum heat sink for any circuit and environment. Most of these formulas consider the temperature and velocity of the ambient air, the amount of heat that must be dissipated, the barometric pressure, and other factors, to determine the heat sink dimensions, composition, number of fins, and required thermal efficiency.

In addition to considering theoretical calculations, practical economic and physical considerations are important in selecting a heat sink. For example, although beryllium copper has a higher thermal conductivity than aluminum, it's generally too expensive to use as heat sink material in CF lamps. Similarly, a massive extruded aluminum heat sink might provide the best protection against thermal overload, but the space limitations imposed by the CF base demand more compact heat sink designs. There's also the issue of availability. If you check parts suppliers for bolt-on heat sinks, you'll find choices for TO-220 devices are limited to aluminum designs rated from 1W to 2.5W. In the end, you go with what you can afford and what fits in the space allotted to the heat sink.

Toroidal Transformer

If you've worked with switching power supplies, the toroidal transformer shown in Figure 6-6 should look familiar. This high-frequency transformer, which, according to my measurements operates at 50kHz, has three windings. Each winding consists of two turns of enameled copper wire. The plastic base provides support and insulation from the underlying circuit board.

FIGURE 6-6 Toroidal transformer

Note the two transistors on either side of the transformer. The unbanded DIAC is visible in the background.

Inductors

This circuit has two inductors: a 100µH ferrite core inductor used to limit EMI induced into the 120VAC line (see Figure 6-7a), and a much larger 2.1mH iron-laminate core inductor used to limit bulb current once the lamp has been started (see Figure 6-7b).

Capacitors

At the center of the disc are two 47µf electrolytic capacitors rated at 250VDC. The remaining capacitors are silver mica rated between 400 and 1250VDC. Figure 6-8 shows several 0.1µf at 400VDC silver mica capacitors in the foreground and the two radial lead 47µf electrolytic capacitors in the background. Silver mica capacitors are a good choice for this application because of their stability over a wide range of temperatures and availability with high working voltages. Note the diodes adjacent to the capacitor leads in the figure.

Resistors

Several generic 0.25W metal film resistors are in this circuit. Of note, however, is the 0.47Ω fusible metal film resistor in series with the 120VAC mains. This resistor is designed to fail flamelessly in the event of other component failure.

a

b

FIGURE 6-7 Ferrite and iron-laminate core inductors

FIGURE 6-8 Silver mica (foreground) and electrolytic capacitors

I like to think of fusible resistors as ultra-slow-blow fuses. It might take a minute or more for the resistor to fail—more than enough time for a mosquito that falls across a capacitor's leads to be vaporized, for example. This resistor is designed to save your house from burning down, should your CF lamp fail.

Diodes

As illustrated in Figures 6-4 and 6-8, several diodes are included in this CF circuit: two for the main rectifier, two associated with the lamp, and five associated with the transistors. Two 1N5397 diodes are part of a voltage doubler rectifier circuit. These diodes have a forward current capacity of 1.5A and PIV (peak inverse voltage) rating of 600V.

Each transistor is protected from peak reverse voltages by a pair of BYV26 diodes. These are ultrafast recovery diodes with surge current capacity of 30A and a PIV of 600V. There is a diode from base to emitter and from collector to emitter of each transistor. The diodes essentially prevent the base-emitter or collector-emitter junctions from being reverse-biased by a potentially damaging transient.

A fast recovery 1N4937 diode is used across each of the filament cathodes of the fluorescent bulb. The polarities of the diodes are such that when one diode is forward-biased, the other is reverse-biased. I assume this design is intended to extend the filament lifespan of the bulb. Can you provide another explanation?

A 1N4005 generic silicon power diode rated at 1A and 600V PIV is used in the oscillator circuitry associated with one of the transistors. More intriguing, however, is a DB3C531 DIAC that is connected to the base of one of the transistors. A DIAC is essentially a momentary latching switch formed by the equivalent of two Shockley diodes wired in parallel, cathode to anode. Once a DIAC's firing voltage is reached, it latches on until the voltage drops below a level sufficient to keep the device conducting. With a sinusoidal AC signal, this cyclical firing, latching, and low-current dropout occurs twice per cycle.

How It Works

Figure 6-9 provides a high-level view of CF lamp operation. A line-powered AC-to-DC converter feeds a DC-to-high frequency AC converter, commonly referred to as an *electronic ballast*. The function of the ballast is to generate a high voltage across the bulb to ignite or fire the CF lamp and then reduce the voltage and regulate the current through the lamp for the duration of operation.

The key components in Figure 6-9 are the capacitor in parallel with the fluorescent tube, labeled *CP*, and the inductor and capacitor in series with the lamp, labeled *LS* and *CS*, respectively. While all three components form part of the self-oscillating electronic ballast, the primary role of CP is to build up voltage to fire the fluorescent bulb. The series inductor, *LS*, regulates the flow of current once the bulb starts conducting.

The series capacitor, *CS*, is used to block DC to the bulb and insure that only AC contributes to current flow. Operating a fluorescent bulb with DC causes *cataphoresis*, the migration of mercury vapor and other gases to the end regions of the bulb where they do not to contribute to illumination. The value CS is significantly larger than CP, so CS does not interfere with the buildup of voltage across CP.

Let's plug in some component values and explore the circuit. Assume an operating frequency of 50kHz, a CP of 5.8 nanofarad (nF), a total filament resistance

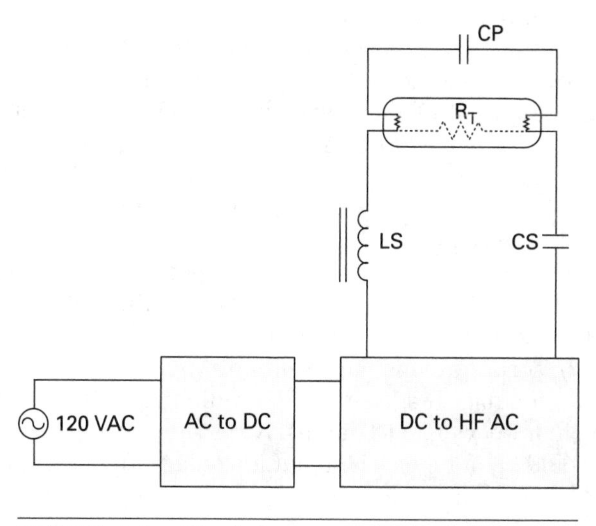

FIGURE 6-9 Block diagram of the CF lamp

of 3Ω, an LS of 2.1mH, and a CS of 0.1μf. When the HF AC is applied, current flows through LS, one filament of the lamp, CP, through the other filament of the lamp, and CS. CP has the following impedance:

$$Xc = 1/(2\pi fc)$$

$$Xc = 1/(2\pi \times 50 \times 10^3 \times 5.8 \times 10^{-9})$$

$$Xc = 550\Omega$$

Using the same formula, you should be able to determine that the impedance of CS is only 15Ω. As such, the majority of the voltage drop will be across CP.

Prior to firing, the resistance of the bulb, RB, is several megaohms, measured from filament cathode to filament cathode. With each cycle, voltage builds up across CP and the bulb. Simultaneously, the filaments decrease the firing voltage by warming and contributing ions to the mercury vapor inside the bulb. At around 500V, the bulb fires, and the resistance of the bulb, RB, drops to about 100Ω, essentially shunting CP and removing it from the circuit. Next, LS and CS, together with the other components in the electronic ballast, control the frequency and magnitude of current through the bulb.

I shorted the filaments and measured the voltage across CP and the bulb. Voltage across the bulb rose exponentially from a few volts to about 600VAC over a period of 1.5 seconds. This cycle repeated dozens of times before I disconnected the circuit to prevent overheating of the transistors.

For steady state operation, the bulb requires about 220VAC from cathode filament to cathode filament. The reason for this relatively low voltage is the operating frequency. At 50kHz, the gas in the bulb remains virtually fully ionized from cycle to cycle. If the bulb were operated at 60Hz, the gas would deionize between cycles and a higher voltage would be required to reionize the gas. As noted earlier, another benefit of the 50kHz operating frequency is the relatively small LS required to limit current flow and prevent the lamp from self-destructing.

Now let's examine the power supply in detail. As shown in the simplified schematic in Figure 6-10, the power supply includes a simple voltage doubler. The 120VAC line is connected to a 0.47Ω fusible resistor and an EMI filter composed of the 100μH ferrite core inductor and two 0.1μf at 400VDC silver mica capacitors. The two 1N5397 silicon power diodes, D1 and D2, charge the two 47μf electrolytic capacitors that are connected in series.

When the hot lead of the applied AC is positive relative to the neutral lead, D1 conducts, charging C1 approximately to the line voltage. When the hot lead is negative relative to the neutral lead, D2 conducts, charging C2. In this way, the two capacitors in series provide the CF circuitry with 240VDC.

A voltage doubler is a low-cost, low-bulk means of providing the higher voltage needed by more powerful CF lamps. A step-up transformer followed by a bridge rectifier would be more efficient, but it wouldn't fit on the circuit board. Lower powered lamps often employ a standard four-diode bridge rectifier and operate at 120VDC.

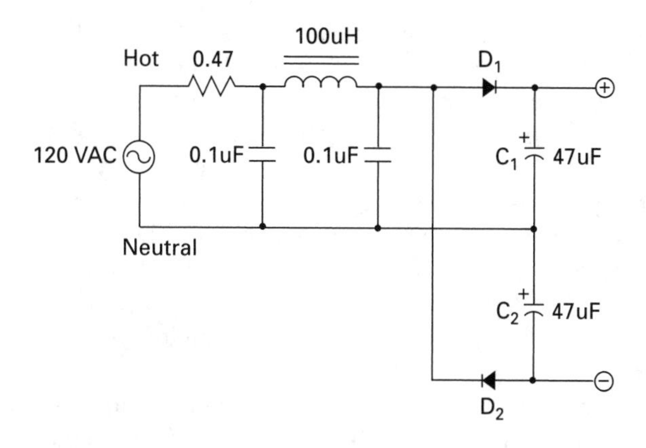

FIGURE 6-10 Simplified schematic of power supply

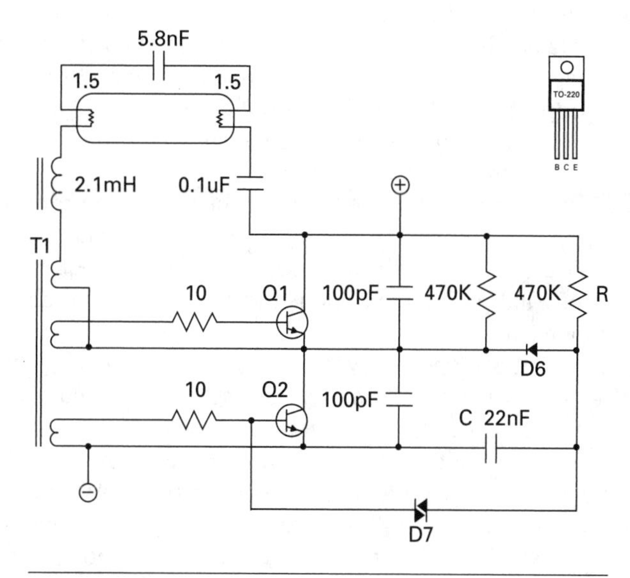

FIGURE 6-11 Simplified schematic of CF lamp

Figure 6-11 shows the simplified schematic of the CF lamp, less the power supply. As you can see, each of the two NPN transistors is connected to a winding of the toroidal transformer T1, and the third winding is connected in series with the bulb circuit. The DIAC is connected to the base of Q2. Note the TO-220 package outline with pinouts for Q1 and Q2.

Despite the relatively low component count, a lot is going on in the circuit that isn't obvious from inspection of the schematic. True analysis requires a multiple-trace oscilloscope and intimate knowledge of the components, especially the transformer, T1, and the transistors, Q1 and Q2. However, at a high level, the manufacturer describes circuit operation as something like the following.

At the moment the power switch is thrown to the on position, the 470KΩ resistor R1, the 22nF capacitor on the emitter of Q2, and the DIAC D7 supply base current to Q2. Q1 is off. Once the bulb fires, the 1N4005 diode D6 inhibits the DIAC. Self-oscillation is maintained by feedback from the base lead of Q2 through the transformer T1. The toroidal core of T1 is formulated such that it easily saturates, thereby removing feedback from the base of Q2, which switches off after a delay defined by the storage time. Meanwhile, T1 switches on Q1, and the circuit oscillation continues.

To get a better handle on this explanation, it helps to know that transistor storage time, or more properly base storage time, is a characteristic of bipolar transistors that limits the maximum speed at which a transistor can be turned off. A charge is stored in the base-collector region of the transistor whenever the transistor is driven to saturation, and this charge takes time to discharge. Q1 and Q2 are both driven to saturation in this circuit.

The 50kHz oscillation frequency is established by the saturation properties of T1 and the transistor storage time—factors also not apparent from the schematic. The saturation properties of T1 are determined by the composition of the toroidal core. The more readily the core saturates, the faster the transistors can switch on and off, within the limits imposed by storage time.

There is a bit of magic in the circuit. The dynamic properties of T1 and storage time don't appear in the schematic, and without these details, it's impossible to specify the changes exactly from startup to steady state. However, I'll offer my impression of what probably happens in the circuit to supplement the condensed version supplied so far.

At the moment the power switch is turned on, the leads of Q1 are at the same potential—the supply voltage—and so the transistor is inactive. In contrast, the emitter of Q2 is at ground, the collector is at positive supply voltage, and the base is forward-biased. In addition, capacitor C charges, and the DIAC fires and latches on. Q2 continues to receive forward-bias from the DIAC and 10Ω resistor in the base lead to ground. Due to T1 saturation, Q1 begins to conduct and has a path to ground through Q2, which is turning off, but not yet off because of storage time. As I see it, it's the slight overlap in on–off time that makes the self-oscillation possible. Otherwise, if Q2 suddenly and completely shuts off, Q1 can't conduct.

Mods

Because a CF bulb is a household appliance, I can't recommend any safe mods if you intend to put the bulb back in regular service. As a bench experiment, you might want to try altering the self-resonant frequency of the CF circuit and record the changes in bulb current, for example. A more appropriate use of

CF lamps, especially those otherwise destined for disposal, is to repurpose the components.

If you search the Web you can easily find examples of a variety of circuits developed with repurposed CF components. As an amateur radio operator (callsign NU1N), one of the more creative ideas I found is to repurpose the components found in a CF lamp to create a radio transmitter. You'll have to rewind T1 and add a crystal to establish the operating frequency. However, other than a DC power supply, everything else can be scavenged from a CF lamp. Coming up with a complete receiver with repurposed components would be more of a challenge.

Chapter 7

Ultrasonic Humidifier

I live in the US Northeast, where the relative humidity, a measure of the amount of water vapor suspended in the air, routinely dips below 20 percent in the winter. In this environment, electrostatic discharges are an everyday annoyance and a serious threat to sensitive electronic components. The "dry air" also makes it tough to sleep at night. A popular way to avoid this scenario is to use a room humidifier, such as the ultrasonic humidifier shown in Figure 7-1, to raise the relative humidity. A value of around 50 percent is comfortable and suppresses electrostatic buildup.

In this chapter, I'll tear down a Crane ultrasonic humidifier, which is typical of the inexpensive room humidifiers available online for about $30. I'll discuss some ideas for repurposing the components in a mod at the end of the chapter.

Highlights

The humidifier consists of a decorative water container supported by a base filled with electronics. The unit features dual linear power supplies, two transistor-based switching circuits, and a high-voltage radiofrequency (RF) oscillator that drives a submerged piezoelectric transducer at about 1.2MHz. The humidifier also uses a brushless DC motor that we'll examine in detail. During the teardown, note the following:

- The open loop control system
- The safety features, including dual fuses and magnetic reed cutoff switch
- The process of circuit analysis

Specifications

Following are the key technical specifications listed by the manufacturer:

- 32W power requirement
- One gallon capacity with 11-hour runtime

FIGURE 7-1 Crane ultrasonic humidifier

- Auto off function
- Variable humidity control

It's a stretch to claim that the unit has a variable humidity control. The single control knob simply varies the excitation of the ultrasonic transducer and therefore the vapor output. As you'll see during the teardown, the unit has no humidity sensor and no humidity feedback mechanism.

Operation

This is a plug-and-play device. Fill the tank with water, place it on the base, toggle the on–off switch to the on position, and adjust mist output with the knob at the base of the unit. A bicolor LED in the base of the unit glows green as long as there is enough water to operate the ultrasonic transducer. When water level above the transducer falls below a few centimeters, a floating magnet drops, causing a magnetic reed switch to open, which kills the ultrasonic transducer and fan. This also switches the LED from green to red, indicating that it's time to refill the tank.

Teardown

You should be able to complete the basic teardown, illustrated in Figure 7-2, in about 15 minutes if you're working with the same humidifier. Most ultrasonic humidifiers on the market share the basic components of this humidifier—a power supply, fan, output control, and ultrasound transducer. From a control perspective, this is an open loop system, in that the power sent to the ultrasonic element is independent of the ambient relative humidity.

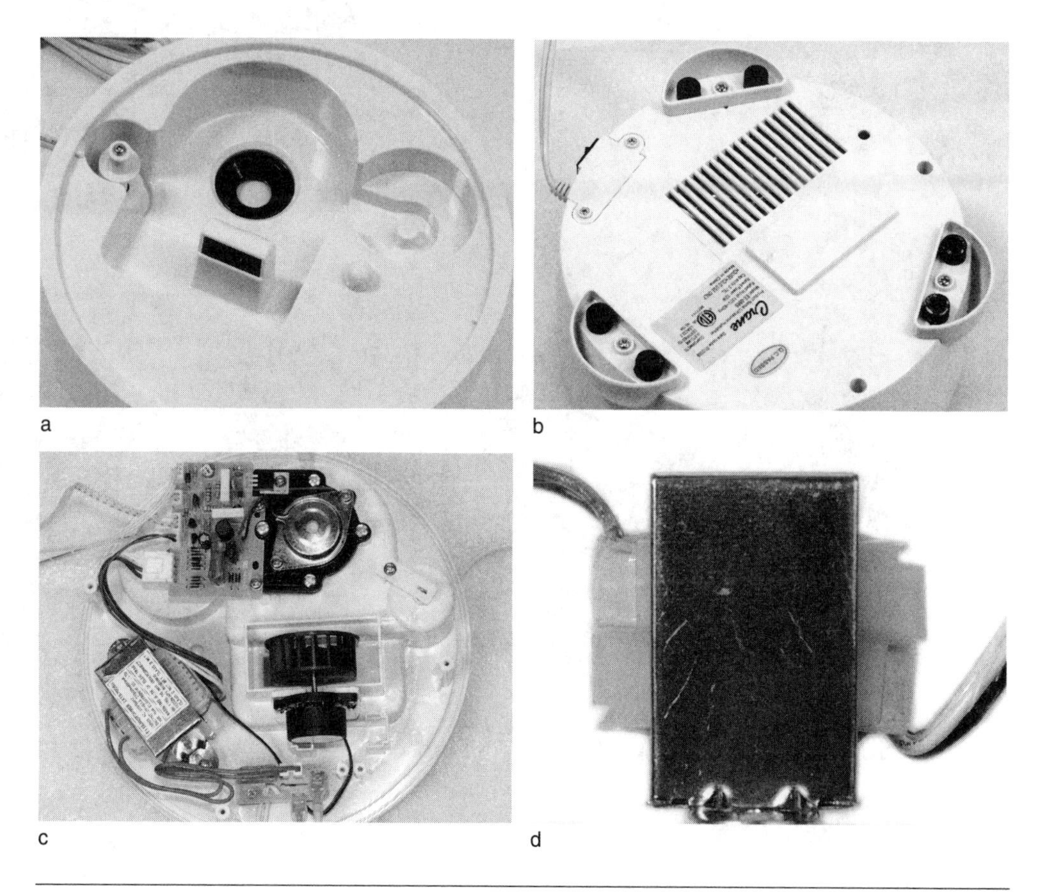

FIGURE 7-2 Teardown sequence

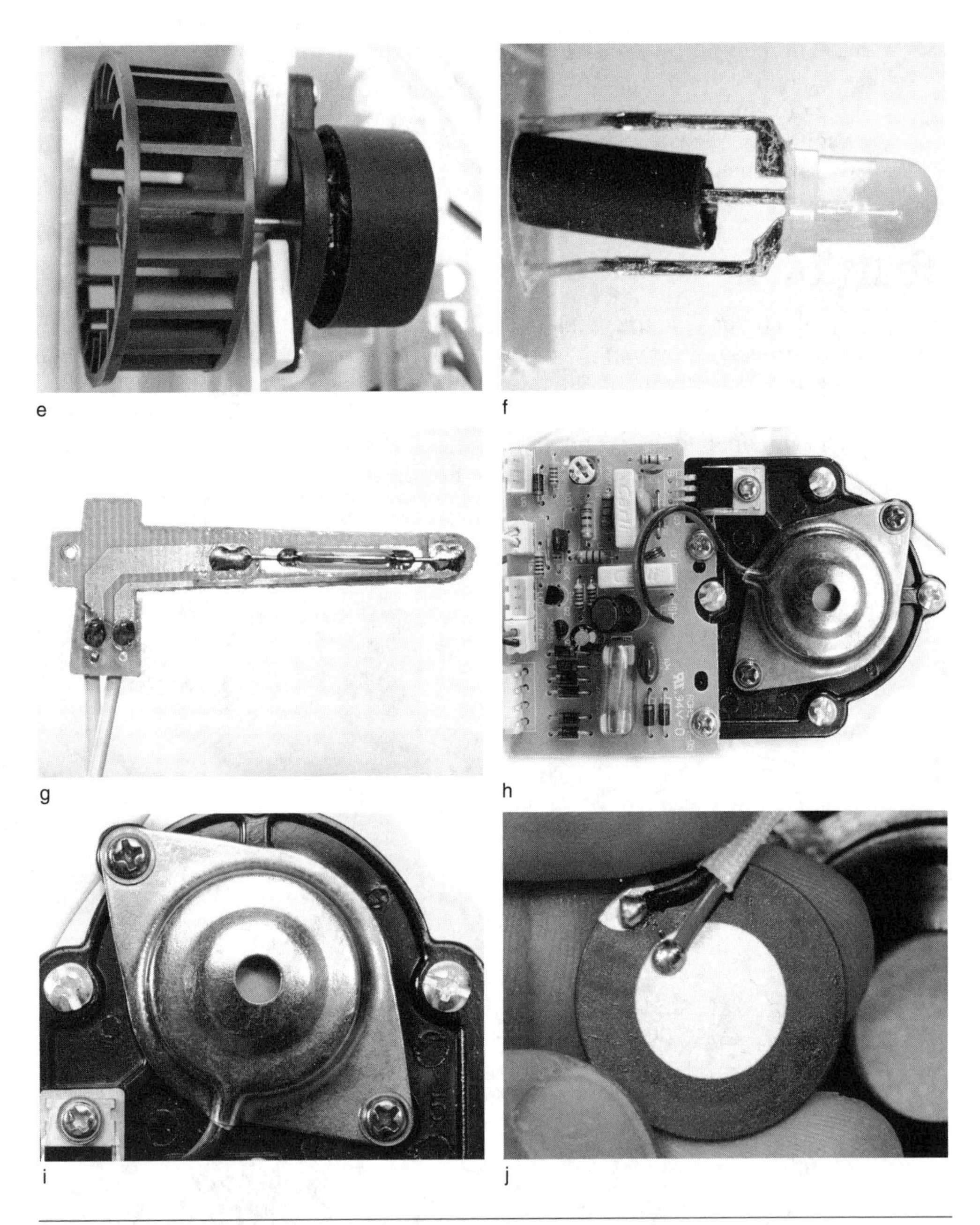

FIGURE 7-2 (*continued*) Teardown sequence

If you're following along at home with another model of humidifier, your unit may have a relative humidity sensor that is used to control mist output. I'll discuss the implications of a closed system operation later in this chapter.

At this point, it's important to note that, despite its appearance, this electric appliance isn't a toy. Not only are we dealing with the potentially dangerous combination of water and electricity, but the ultrasonic transducer can hurt you. The transducer in a typical home ultrasonic humidifier delivers up to 15W—and that's a lot of power at ultrasound frequencies. As a point of reference, the output of therapeutic ultrasound machines is measured in milliwatts.

Before you think of touching the active transducer in your humidifier, keep in mind that this device is designed to shred water at the molecular level, and remember that your finger is mainly water. Although I've not experienced the jolt from an ultrasonic transducer firsthand, I'm told that touching an active transducer is akin to having your finger slammed in a car door. And that, from personal experience, is something to avoid. I'm not suggesting that you pass up the opportunity to repurpose the components to create an ultrasonic meat tenderizer. Just don't touch the active transducer.

Tools and Instruments

You'll need a Phillips-head screwdriver, crescent wrench, and needle-nose pliers. A multimeter capable of measuring capacitance and inductance modes will come in handy.

Step by Step

If you plan to rebuild your humidifier once you're done with the teardown, sort the screws and other hardware in a cupcake pan or similar compartmentalized container. Use a permanent marker to identify connectors before you disconnect them. And pay particular attention to the integrity of the seal around the ultrasonic transducer—a damaged seal will allow water into the electronics compartment.

Unplug the unit before you begin.

Step 1

Examine the base. Remove the decorative water tank, leaving the base, shown in Figure 7-2a. Thoroughly dry the base with a paper towel. With the output control knob facing you, identify the depressed area with the black-rimmed white disc toward the rear of the base. This disc, about the size of a quarter, is the ultrasonic transducer. Toward the left of the transducer, near the left rim of the base, is the magnetic float assembly. When the water level rises, a ring magnet embedded in a buoyant plastic ring moves with the water level and interacts with a magnetic reed switch on the dry underside of the base. The rectangular fan outlet appears just below the transducer in the figure.

Step 2

Crack the base. Flip the base upside down to reveal the fan input grill and screws securing the feet and base, as shown in Figure 7-2b. Remove the screws and lift the bottom of the base to reveal the electronics.

Arrange the base—actually the underside of the base top, which separates the water from the electronics—as shown in Figure 7-2c. At the top of the figure are the main printed circuit board and ultrasonic transducer. Connectors and cables for sensors and controllers run along the left edge of the circuit board. Near the right rim of the base is the magnetic reed switch used to determine water level. Near the center of the base is the 12VDC brushless squirrel cage fan. At the bottom of the figure is a board for the switched AC input and fuse. In the left lower quadrant of the base is the dual-secondary power transformer.

Step 3

Examine the live circuit. Plug in the unit and measure transformer secondary voltages and examine the logic of the magnetic reed switch. If you have an oscilloscope or frequency meter, measure the operating frequency of the ultrasound generator. However, note that the ultrasonic transducer is designed to operate submerged. If you want to measure voltages and currents with a dry transducer, manipulate the magnetic ring so that power is applied to the ultrasound circuitry in 1- or 2-second bursts—just long enough to get solid measurements. You can also unsolder the transducer to avoid overheating and overstressing the component.

Unplug the unit when you're done.

Step 4

Extract the small circuit board. Remove the small board that holds the 2A at 250V fast-blow fuse and that provides quick-disconnect connections to the power cord and the primary of the power transformer.

Step 5

Extract the dual-secondary power transformer, which provides 28VAC at 800mA and 13.5VAC at 90mA secondary output voltages. Note the construction with the insulated coils for the primary winding on the top half of the transformer and secondary windings on the bottom half, as shown in Figure 7-2d. The transformer is virtually waterproof, thanks to a thick layer of clear sealant on the windings and laminated steel core. As with the small printed circuit board, the transformer is physically isolated from other components in the plastic base of the humidifier.

Step 6

Release the brushless DC fan. The squirrel cage fan blades, shown in Figure 7-2e, are pressure-fit on the motor shaft. Simply pull the cage along the axis of the shaft. Without disconnecting the power cable, remove the two screws securing the motor to the plastic support and set the motor down in the base for now.

Step 7

Extract the bicolor LED and 5KΩ potentiometer from the front of the base. You'll need to use a crescent wrench or nut driver to remove the retaining nut from the potentiometer. Figure 7-2f shows the three-lead, common cathode bicolor LED. Don't disconnect the potentiometer cable from the main circuit board.

Step 8

Extract the magnetic reed switch assembly. Partially remove the screw securing the plastic pressure tab and rotate the tab off of the magnetic reed switch assembly, shown in Figure 7-2g. Pull the assembly straight up, out of its well in the base. You may find it easier to extract the sensor if you temporarily unplug its cable from the main ultrasonic transducer board.

Step 9

Extract the main ultrasonic transducer assembly. Carefully remove the four Phillips-head screws that attach the large aluminum heat sink to the plastic base, as shown in Figure 7-2h. The assembly includes a heat sink about the same size as the main circuit board. Note the clear silicon ring seal between the plastic base and heat sink.

Step 10

Extract the piezoelectric transducer. Remove the two screws securing the silver metal retainer to the black heat sink, shown in Figure 7-2i. Peel the black rubber seal from the piezoelectric disc, taking care not to damage the two wires connecting the transducer to the main printed circuit board. The free transducer is shown in Figure 7-2j.

Layout

The layout of the main circuit board is clean and relatively spacious; you shouldn't have any trouble probing component leads. As a bonus, the component side of the board, shown in Figure 7-3, is marked with component values. As you can see, the board contains traditional leaded components. The connectors along the bottom edge of the board, from left to right, are for the control potentiometer (VR), the magnetic reed switch (K), the bicolor LED (LED), the fan (FAN), and power (C).

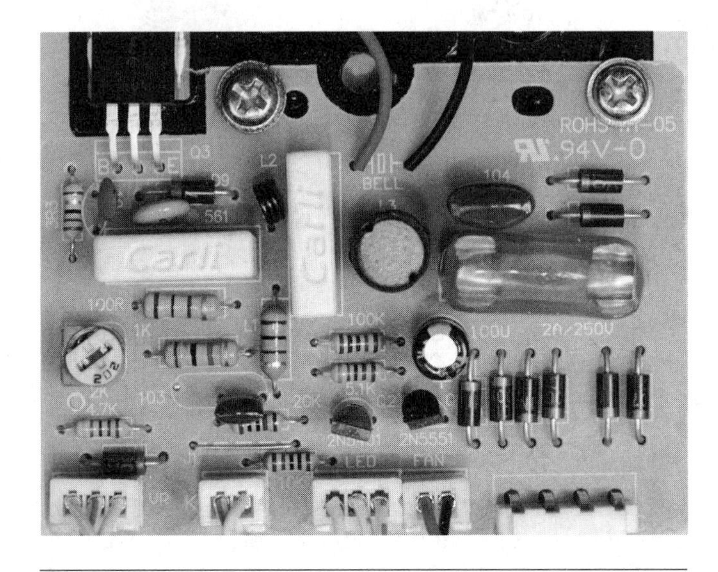

FIGURE 7-3 Main circuit board layout

The right side of the board is populated with components for the two power supplies—eight standard silicon power diodes configured as two bridge rectifiers, one for the 28V and one for the 13.5V transformer outputs. The 100µf at 25VDC electrolytic capacitor (100U) is a filter capacitor in the 13.5V circuit, and the 2A at 250V fast-blow fuse (2A/250V) is in the 28VDC circuit that feeds a high-voltage oscillator. The 0.1µf Mylar film capacitor (104) is used as a bypass capacitor in the power supply.

Moving to the left, immediately adjacent to the fan and LED connectors, are the small switching transistors in TO-92 packages, a 2N551 NPN (negative-positive-negative) silicon transistor for controlling the fan motor and a 2N5401 PNP (positive-negative-positive) silicon transistor for controlling the red element of the bicolor LED.

On the left side of the board, behind the connectors for the magnetic reed switch and control pot, is the drive circuitry for the ultrasonic oscillator. The 2KΩ trim pot (potentiometer) (2KΩ) is wired in series with the external 5KΩ control potentiometer. The oscillator transistor, a BU406 high-voltage NPN silicon transistor in a TO-220 case, is bolted to the large heat sink. You can see a small amount of conductive grease between the transistor's metal tab and the large aluminum heat sink. The transistor's metal tab, which is connected to the collector, is at 28VDC during operation of the humidifier—something to keep in mind if you're probing the live circuit.

Other significant components in the area include an open-air RF inductor composed of three loops of enamel wire (L2), a cylindrical RF inductor (L3) adjacent to the electrolytic filter capacitor, and a 100µH inductor (L1). Adjacent to the open-air RF inductor are two ceramic disc capacitors (101 and 561) and two large, rectangular metalized polypropylene film capacitors.

Components

This simple humidifier provides examples of a variety of components that you'll see in other appliances and in other teardowns in this book. When you discover components during a teardown, make a habit of looking up the datasheets on the Web. Datasheets often feature example circuit diagrams and other information that may not be evident in the teardown.

Capacitors

As discussed earlier, the humidifier employs electrolytic, Mylar film; metalized polypropylene film; and ceramic disc capacitors. The two Mylar film capacitors, marked "2A103J" (0.01µf) and "2A104J" (0.1µf), are perfect for high-humidity environments because they are dipped in a hard epoxy coating for moisture resistance. The small ceramic disc capacitor, marked "101" (100pF), is installed between the base and collector of the oscillator transistor. The large, resin coated disc capacitor, labeled "561K 2KV" (560pF), has a 2KV voltage rating.

Piezoelectric Transducer

The heart of the humidifier is also one of the simplest components, the piezoelectric transducer, which imparts ultrasound energy from the oscillator circuit to the water. There are no markings on the transducer in my humidifier. However, I'd expect the datasheet to include information on the resonant frequency, static capacitance, mist production in cc/minute, voltage input, and maximum drive capacity. The voltage across the disc during operation is more than 120V peak-to-peak (VPP).

Diodes

The ten diodes used in the humidifier are general-purpose 1N4007 silicon diodes rated at 1A and 1000V PIV (peak inverse voltage). Two sets of four diodes are used in bridge rectifier circuits: one is used to protect the high-voltage oscillator, and the other isolates the oscillator circuit from the other circuits on the board.

Bicolor Red/Green LED

The bicolor LED used to signal the status of the humidifier has a common cathode center lead. The LED emits red light when the anode corresponding to the red LED is positive, and it emits green when the other anode is positive.

Magnetic Reed Switch

The slender magnetic reed switch is normally open. In this application, when there is sufficient water in the base, the floating ring magnet keeps the reed switch closed. If the water level drops sufficiently, the ring magnet moves away from the reed switch, allowing it to open.

SPST Rocker Switch

The SPST (single pole, single throw) rocker switch is rated at 6A at 125VAC. If you repurpose this switch for a DC application, remember to derate the current capacity rating by at least 50 percent.

RF Inductors

In addition to L1, the banded 100μH inductor (brown-black-brown-silver), are two unmarked inductors. We can estimate the value of L2, the air-core coil, with the aid of the following formula:

$$L\ (\mu H) = d^2n^2/(18d + 40ln)$$

where d = coil diameter in inches, from wire center to wire center; ln = length of coil in inches; and n = number of turns. My measurements for the RF inductor are as follows: d = 0.16 inch; ln = 0.08 inch; and n = 3. Substituting these measurements into the formula, we have this:

$$L = 0.16^2 3^2/(18 \times 0.16 + 40 \times 0.08)\mu H$$

$$L = 0.23/6.1 = 0.038\mu H$$

Estimating the value of L3 is more problematic because of the unknown permeability of the ferrite core. Ferrite can provide 20–1000 times more inductance per turn than an open-air inductor of the same size. According to my measurements with an LC meter, the inductance of L3 is 32μH.

Heat Sink

The 1.1 ounce (31g), 2.5 × 2.6 inch (65 × 67mm), aluminum heat sink seems more than adequate to cool the oscillator transistor, even with continuous, day-after-day operation. Given the mass of the heat sink and its contact with water, it provides more than sufficient heat capacity for the transistor. The ultrasound transducer is insulated thermally and electrically from the heat sink by a silicone seal.

Resistors

The main circuit board contains a handful of 0.25W metal film resistors; a 0.1W, 2KΩ trim pot; and an external 0.5W, 5KΩ potentiometer. Of note is the similarity of the 100Ω resistor and 100μH inductor. The two components are about the same size and share a color code (brown-black-brown-silver), but the resistor has capped ends and the inductor has conical ends.

Transistors

As noted earlier, three silicon transistors are used in the humidifier, a 2N551 NPN and 2N5401 PNP low-power switching transistor and a hefty BU406 NPN high-voltage transistor. The datasheets for the transistors reveal a current gain (Hfe) of 50–250 for the 2N551 and 2N5401 and 20–40 for the BU406. Let's review the on and off states for NPN and PNP bipolar transistors used as switches, using Figure 7-4 as a guide.

Recall that an NPN bipolar transistor can be configured as a switch between emitter and collector by manipulating the emitter-base current. When the base is positive relative to the emitter, a small current flows from emitter to base and a significantly larger current flows from emitter to collector. The amount of emitter-base current we need to switch a particular load is related to the load and the transistor current gain. With a current gain of 100 and a load of 100mA in the emitter-collector circuit, we'd need an emitter-base current of 100mA/100, or 1mA. Similarly, with a current gain of only 50, we'd need an emitter-base current of 100mA/50, or 2mA, to switch the 100mA load on and off reliably.

The BU406, which is capable of switching 10A at 200V and dissipating 60W, has the bias polarity requirements as the diminutive 2N551. Given a current gain

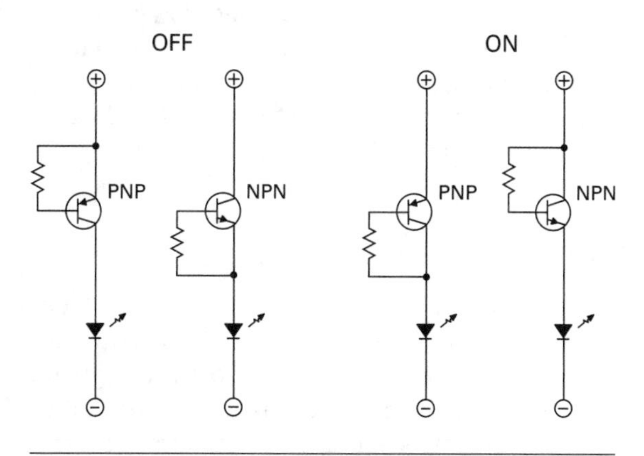

FIGURE 7-4 NPN and PNP transistor switch on and off conditions

of 30 and a load of 10A, we'd need an emitter-base current of 10/30, or 300mA, to reliably switch the 10A load on and off.

A PNP bipolar transistor can also be configured to switch a load in the emitter-collector circuit by manipulating the emitter-base current, as long as the proper bias polarities are observed. When the base is negative relative to the emitter, a small current flows from emitter to base and a significantly larger current flows from emitter to collector. As with the example of the NPN transistor, the transistor's current gain determines the emitter-base current required to switch the expected emitter-collector current reliably. With a current gain of 100 and a load of 20mA, the emitter-base current should be 20mA/100, or 0.20mA. Similarly, with a current gain of 50, the emitter-base current should be 20mA/50, or 0.40mA.

Compared to the diminutive 2N551 and 2N5401 transistors, the BU406 is a powerhouse capable of switching 10A at 200V and dissipating 60W.

Fuses

Both fuses in the humidifier are 2A fast-blow designs. Whereas many appliances feature slow-blow fuses, the fast-blow variety was selected apparently because of the need to kill the power quickly in the event of water entering the electronics area. Even so, the fuses seem overly conservative. At 32W, the humidifier should draw a little more than a quarter amp:

$$I = P/V = 32W/120V = 270mA$$

A 1A fast-blow fuse is probably a better option. One reason for using a 2A fuse would be to allow for power-on surges. However, in my tests, I didn't detect any surges that would blow a 1A fuse.

You can usually differentiate fast- and slow-blow fuses by the shape and size of the fuse element. Fast-blow fuses are constructed with a single thin wire from end cap to end cap. Slow-blow fuses, in contrast, are usually made of a flat, wide conductor that narrows slightly toward the middle of the fuse body. When in doubt, read the end caps.

Power Transformer

The step-down power transformer has a laminated steel core with a 120V primary and two secondaries, one rated at 28V at 80mA and one at 13.5V at 90mA.
An individual lamina from the core is shown in Figure 7-5a. The steel core is responsible for efficient coupling of magnetic flux induced by the primary winding to the secondary windings. This particular core design, called an *EI core* because the core is formed by a stack of *E*-shaped steel plates, such as the plate shown in Figure 7-5b, is mated with *I*-shaped steel plates. This common design is inefficient but inexpensive to manufacture.

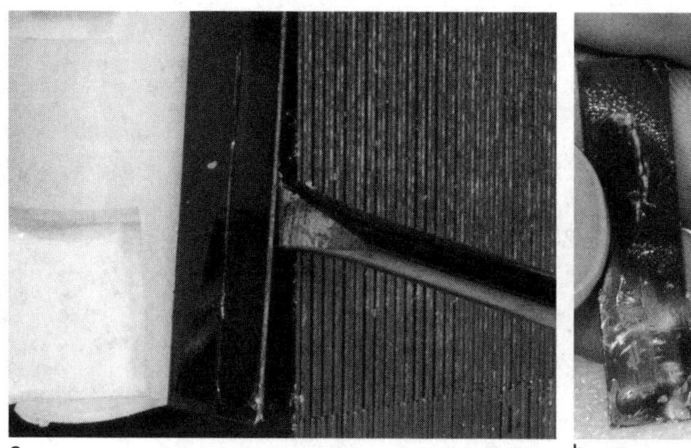

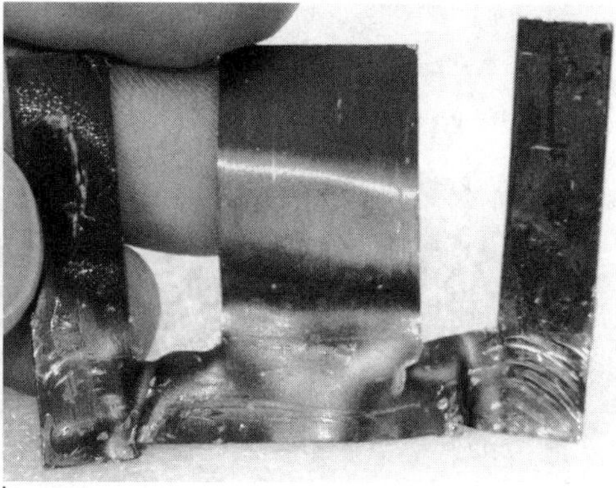

a b

FIGURE 7-5 Iron laminate transformer E plate

Brushless DC Motor and Fan

Let's explore the 12V/80mA DC brushless motor (BLDC) using Figure 7-6 as a guide. Remove the squirrel cage fan with your hands, if you haven't done so yet, and then, with a pair of pliers, twist the shaft counterclockwise, holding the cap steady. The shaft should release from the cap and pass through the center of the coil assembly without resistance.

Figure 7-6a shows the cap containing a DC permanent magnet and the coil assembly. Leads to the 13.5VDC supply are attached to the bottom of the coil assembly. A side view of the coil assembly, shown in Figure 7-6b, reveals the coils of enameled wire and a 4-pin IC flanked by two 3.3µf at 16VDC electrolytic capacitors.

Notice that there are no brushes in the motor. In operation, the coil assembly or stator remains stationary while the DC magnet, which is glued to the inside of the plastic cap, rotates. The only point of contact is the shaft, which isn't strictly necessary as long as the cap can rotate freely around the coil assembly.

Let's take a look at the magnet. It measures about 1 inch (25mm) inside diameter, 1/2 inch (1.6mm) thick, and 3/8 inch (10mm) deep. What isn't clear from visual inspection is the number and orientation of magnetic poles—without instruments. Figure 7-6c shows the four poles of the circular magnet, assessed with the aid of a magnet polarity tester (available from Stewart MacDonald at www.stewmac.com) and magnetic field viewing paper (available from Scientifics at http://scientificsonline.com). Three of the circular magnet's four poles are visible as pie slices in the viewing paper. The polarity tester shows poles alternating between north and south every 90 degrees around the circumference of the magnet. That is, the two north and south poles are on opposite sides of the magnetic band.

FIGURE 7-6 Brushless DC motor construction

If you look closely at Figure 7-6d, you can see that the silver wings on each of the four ferromagnetic cores is slightly tapered on the counterclockwise-most end. At power-down, the resulting air gaps between the rotating magnet and the stationary coils causes the magnetic rotor to stop so that the north–south pole boundaries are aligned with the spaces between ferro cores. That is, the magnet will stop so that the poles line up over the centers of the coils. As discussed in the next section, this simplifies the startup procedure. The magnetic cap spins counterclockwise relative to the coils shown in the figure.

How It Works

At the heart of a home ultrasonic humidifier is an ultrasonic oscillator that drives a piezoelectric transducer. When excited by high-energy ultrasound from the transducer, small globules of water and organic and inorganic matter suspended in the water are launched into the air. A fan creates an air current that transports this water vapor out into the room or other environment. When the water evaporates, these impurities are distributed around the humidifier in the form of an annoying white dust.

With this background, let's explore the inner workings of our ultrasonic humidifier, starting with the power supply. As shown in Figure 7-7, the power supply is a typical linear design with full-wave bridge rectification in both secondaries. The 13.5VDC supply, which supplies the fan motor, has a 100µf filter capacitor. The 28V supply is unfiltered. There is a 2A fuse in the primary and 28VDC secondary. Of note is that the fuse in the secondary is in the ground lead.

As illustrated in the simplified schematic in Figure 7-8, the power supply provides 13VDC to the fan circuit, composed of Q1 and the brushless DC (BLDC) motor. The power supply also provides 28VDC to Q2, which switches the bicolor

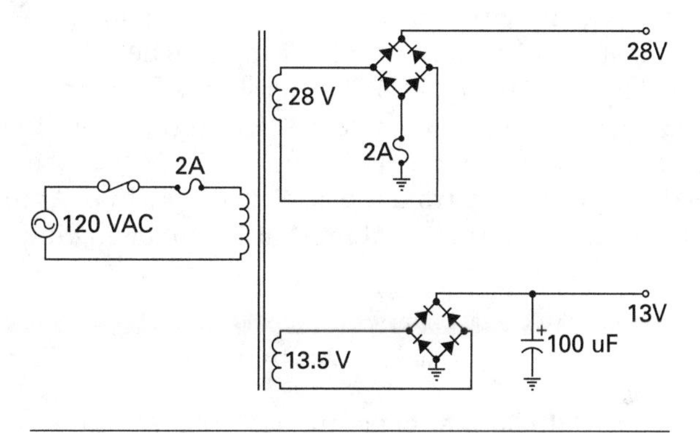

FIGURE 7-7 Schematic of power supply

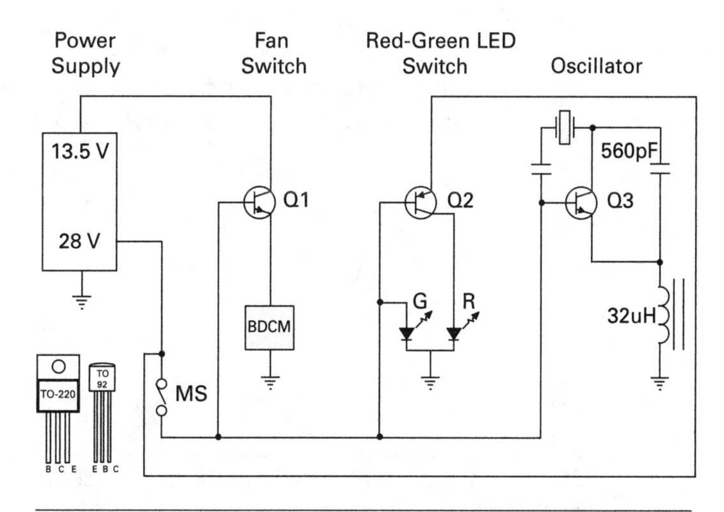

FIGURE 7-8 Simplified schematic of the humidifier

LED, and to Q3, the ultrasonic oscillator. When the magnetic switch (MS) is closed, indicating there is water in the base of the humidifier, the bases of all three transistors are biased positively, turning on Q1 and Q3 and switching off Q2. As a result, the fan and oscillator becomes active, the green LED is illuminated, and the red LED is extinguished. Note that, for clarity, the various bias resistors are not included.

The resonant frequency of the oscillator is established by the series-resonant circuit of the 560pF capacitor and the 32µH inductor. We can calculate the theoretical resonant frequency of the series LC circuit with the following formula:

$$F = 1/(2\pi\ (LC)^{1/2})$$

$$F = 1/(2\pi\ (32 \times 10^{-6} \times 560 \times 10^{-12})^{1/2}) = 1.2\text{MHz}$$

This is in agreement with the actual operating frequency of the oscillator.

Let's look at the ultrasonic oscillator in more detail, using Figure 7-9 as a guide. As discussed earlier, Q3, the BU406 transistor, is forward-biased when the magnetic switch (MS) is closed. The 5KΩ potentiometer establishes the amount of forward bias and therefore the output of the oscillator. The 100µH choke (L1) decouples the RF generated in the oscillator from the 28VDC supply line. At the operating frequency of about 1.2MHz, the choke would present the following impedance:

$$X_L = 2\pi FL$$

$$X_L = 6.2 \times 1.2 \times 10^6 \times 100 \times 10^{-6} = 745\Omega$$

Although the BU406 is a robust transistor, it isn't immune to damage. As noted earlier, I measured 120VPP across the transducer, which is fairly close to the

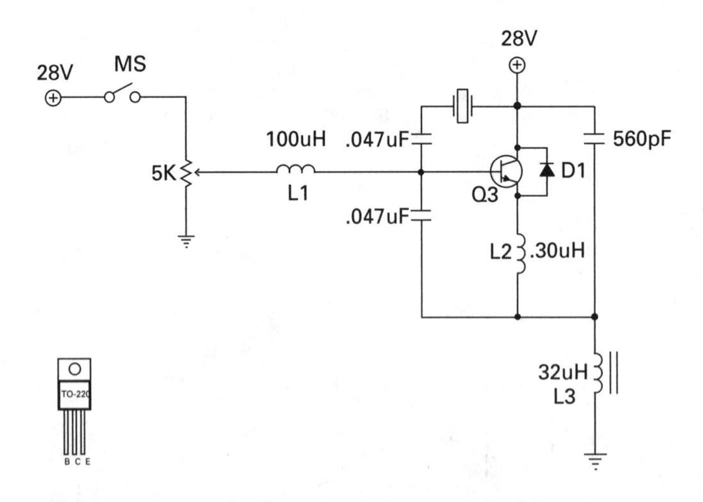

FIGURE 7-9 Simplified schematic of ultrasonic oscillator

transistor's maximum rating of 200V. A diode from the collector-emitter circuit (D1) protects the transistor from reverse overvoltage spikes generated by the crystal transducer. If the collector becomes more negative than the emitter, the diode conducts, shunting the current away from the transistor. The diode has no effect in the amplitude or frequency of the oscillator output.

The role of the 0.30μH open-air inductor (L2) may also be associated with maintaining the integrity of Q1, including inhibiting parasitic oscillations. I am certain it has no bearing on normal operation—the output amplitude and frequency were unaffected when I replaced L2 with a short length of wire.

Now let's turn our attention to the BLDC motor. As noted earlier, a brushless motor is a natural fit for our humidifier because there are no brushes, commutators, slip rings, or other electromechanical components to corrode and malfunction in a humid environment. In addition, there's no need for periodic maintenance, and brushless motors are quieter than their brush counterparts. If you're unfamiliar with the characteristics of a brush motor, think power drill—loud and electromagnetically dirty, but powerful.

Our motor has stationary coils and an outer, rotating plastic cap with a magnetic band glued inside. The motor is fed with 13VDC, with no other electrical input or output. So how does the motor turn? Since the polarity and magnitude of the magnetic field produced by the permanent magnet in our BLDC motor is fixed, the electrical activity necessary for rotation must come from the coils and drive circuitry.

The four-lead controller chip soldered to the BLDC motor circuit board has several potential functions. First, it certainly generates and feeds the coils with an AC signal of the appropriate frequency and amplitude. Second, it may sense the position of the rotating magnet. I couldn't locate the datasheet for the controller chip, so I can't say for certain, but the construction of the motor suggests this may be a sensorless motor design.

The controller chip may detect when magnetic polarity changes as the north-south pole passes immediately in front of it. There are several ways to detect this change in magnetic polarity. One way is to monitor the back electromotive force (EMF) produced by the spinning magnet and stationary coils, just as though the motor were a generator. Another popular approach is to use a Hall effect sensor, which senses magnetic flux density and polarity. It's also possible to go sensorless.

My first impression was that our BLDC motor was sensorless, and that the controller chip functioned solely as an AC generator. To determine whether a Hall effect sensor was buried in the controller chip, I removed the permanent magnet cap and then, with power applied, moved a small magnet across the face of the chip. The magnet triggered a series of trapezoidal AC pulses from the chip—exactly what I'd expect from a chip with a Hall sensor.

However, another factor that suggests sensorless operation is the tapering of the four ferro cores, shown in Figure 7-6d. One of the challenges of working with a BLDC motor in a sensorless configuration is that there is no information on the position of the rotor at startup. Without this information, it's difficult to ensure that the AC signal sent to the coils will result in rotation in the proper direction. One solution to this dilemma is to shape the ferro cores so that they bias the system to rotate in one direction over another. My conclusion is that the BLDC motor uses a combination of a Hall sensor and shaped ferro cores—probably in a redundant configuration, so that one system can take over if the other fails.

Mods

This humidifier is a great source for components that can be repurposed for other projects. I can think of numerous applications for the intact ultrasound module, repackaged so that the transducer is mounted in a probe or in a small container. You could create an ultrasonic cleaner, a mixer for small quantities of fluid, and maybe even a meat tenderizer. Just remember to keep your fingers away from the transducer and don't allow the module to overheat.

Because of safety issues, there isn't much in the way of mods if you plan to use this humidifier unattended in your home on a constant basis. However, this humidifier is a perfect experimental platform for working with closed-system concepts for a science fair project or simply to get a better understanding of the practical application of feedback and control theory.

Closed Loop System

Like most inexpensive humidifiers, our humidifier is an open loop system. You set the rate of vapor production and periodically monitor the status LED for an indication of when it's time to refill the unit. If you happen to forget to fill the reservoir, the unit will shut off automatically. The limitation of this open loop design is that the humidifier output doesn't reflect environmental conditions. Ideally, it would produce more water vapor in dry weather and less in damp, humid weather.

You could form a human-in-the-loop closed loop system by sitting next to the humidifier with a hygrometer in one hand and the humidifier output control in the other. When the humidity drops below your established optimum value, you crank up the output. Conversely, when the humidity exceeds your preset value, you turn down the output. You could automate this closed loop control process by using a humidity sensor, a means of setting the desired humidity, and a difference detector that's linked to the output controller, as shown in Figure 7-10. Keep this in mind when you read the next chapter teardown of a hygro thermometer.

Once you've established a means of detecting when the relative humidity is above or below your set point, you can use a microprocessor to control a relay in series with the magnetic reed switch to activate and deactivate the humidifier at the set points. You'll find that if the high and low humidity control points are too close together, the humidifier will constantly cycle on and off because of the tolerance of the sensor and other components.

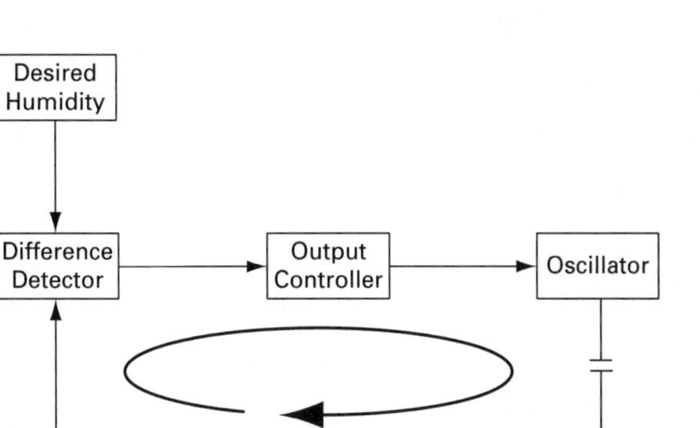

FIGURE 7-10 Closed loop system diagram

The closed loop system would also be affected by temperature and the size and nature of the room. For example, if you put the humidifier in a closet, the humidity will tend to overshoot the set point. If you put the humidifier in a large room, it may take hours for the humidity to change significantly. Open the windows or turn on an air conditioner, and the humidifier may never cycle off or on, depending on set points.

The programming challenge is to control the closed loop so that the unit isn't switching on and off five times a second in response to small deviations in the difference detector, or so that it isn't cycling every 15 minutes, well after the humidity has swung significantly greater or less than the set point. As you might expect, developing a computer algorithm that can maintain an exact relative humidity is more difficult than it sounds.

If you're interested in tackling this challenge, you can find a body of literature and code samples that address closed loop control, including the popular PID (Proportional-Integrative-Derivative) model. The robotics community has also developed practical hardware and software solutions to closed loop control systems that are worth investigating.

Chapter 8

Digital Hygro Thermometer

If you read the preceding chapter describing the teardown of an ultrasonic humidifier, you know the advantage of having a digital instrument that accurately measures the relative humidity as part of a closed loop control system. Add to this capability a digital thermometer and the instrument becomes an indispensable comfort gauge.

I own several inexpensive digital hygro thermometers and find them indispensable. There's one in my guitar room to help me ensure that my wooden instruments don't turn to splinters or develop mildew, one in the bedroom to monitor the creature comfort levels, and a USB hygro thermometer in a server closet to ensure the environment is suitable for the computers, hard drive arrays, and communications devices. The USB unit is wired to my workstation so that I can monitor the environment remotely.

In this chapter, I'll tear down an Extech Hygro-Thermometer, shown in Figure 8-1. This digital meter is sold under a variety of brands for about $40. I'll discuss some ideas for repurposing the components and a mod at the end of the chapter.

Highlights

The most obvious component of our digital hygro thermometer is the large LCD display. The actual measurement and display hardware, tucked away on a circuit board in the lower fifth of the unit, includes a microcontroller, a crystal oscillator, an analog humidity sensor, and an analog temperature sensor. Unfortunately, the mystery microcontroller is mounted directly on the circuit board and buried under a glob of black epoxy.

During the teardown, note the following:

- The use of SMT components
- The ribbon cable connecting the LCD to the circuit board

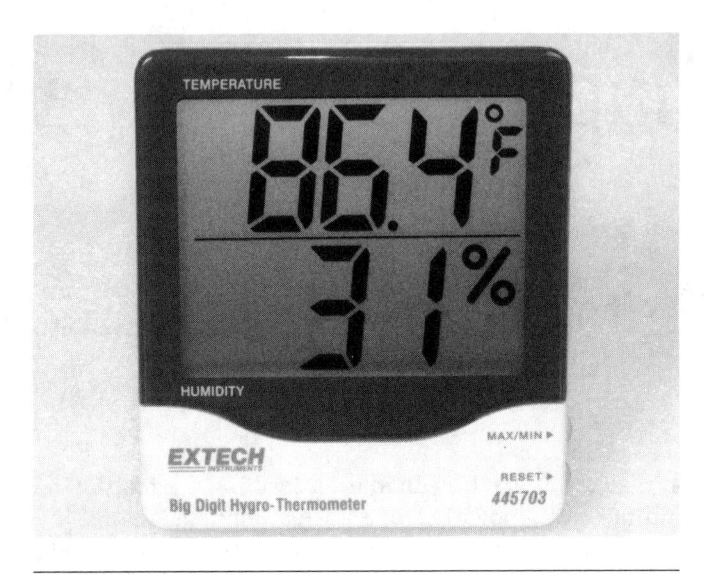

FIGURE 8-1 Digital hygro thermometer

- The crystal oscillator
- The analog temperature and humidity sensors

Specifications

Following are the key specifications of the instrument:

- 4 × 4 × 3/4 inches (HWD)
- Digital display of temperature in degrees Fahrenheit or centigrade (°F or °C) and relative humidity
- Memory function to store and display minimum and maximum temperature and humidity values
- Temperature accuracy is ±1.8°F from 14°F to 122°F (±1°C from –10°C to 50°C)
- Relative humidity accuracy is ±5 percent RH from 25 percent RH to 85 percent RH within the temperature range of 32°F to 122°F (0–50°C).
- Powered by a single 1.5V AAA cell with a six-month life expectancy

These accuracy figures are typical for low-cost digital hygro thermometers. Meteorological digital and analog hygro thermometers capable of measuring relative humidity within ±2 percent of the actual value are available starting at about $200. Of course, laboratory grade instruments with greater accuracy are available.

Operation

Only two buttons and a switch are required to operate this otherwise hands-off meter. A F°/C° switch on the back of the meter determines whether temperature is displayed in Fahrenheit or centigrade, respectively. In the lower-right corner of the front of the meter are the Max/Min and Reset buttons. The Max/Min button recalls the maximum and minimum temperature and relative humidity values since the Reset was pressed. The Reset button erases temperature and humidity values stored in memory.

Teardown

You should be able to complete the basic teardown, illustrated in Figure 8-2, in about 3 minutes. Add another few minutes if you unsolder the sensors.

Tools and Instruments

You'll need a small Phillips-head screwdriver to crack the case and a soldering iron if you want to remove and experiment with the sensors. This is a nice meter for a desk or wall, and you needn't destroy it during the teardown.

Step by Step

Although it's not strictly necessary, you can verify that the AAA cell holder is empty before you begin. You can easily replace the battery once the circuit board is exposed.

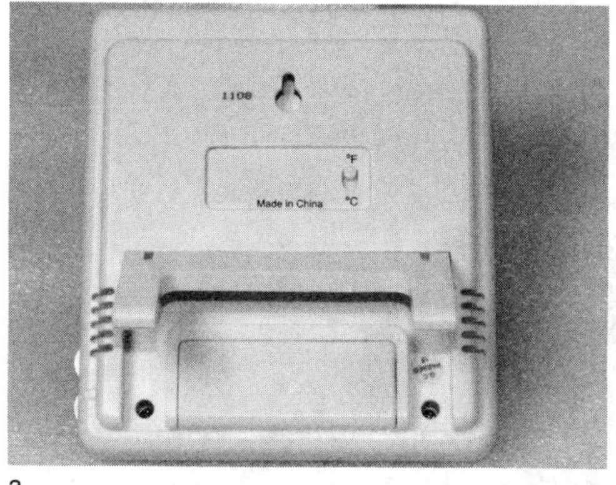

a

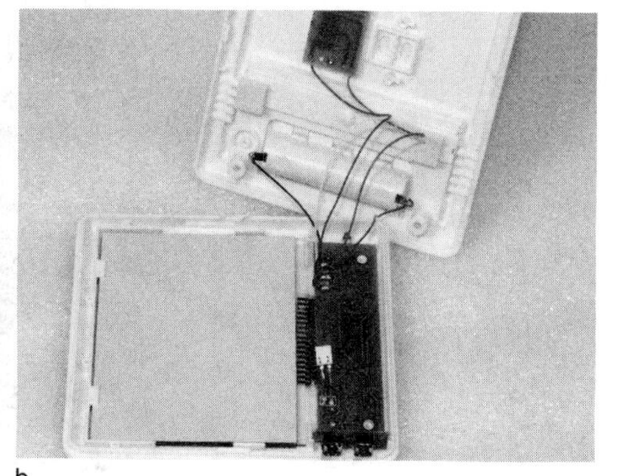

b

FIGURE 8-2 Teardown sequence

Step 1

Disengage the back. With the meter face down, lift up the plastic easel stand to expose the two Phillips-head screws, as shown in Figure 8-2a. Remove the screws. With your fingernails or a coin, push in the four tabs around the edge of the meter—two on top and two on either side. Gently open the back to reveal the LCD and circuit board, shown in Figure 8-2b. Avoid tugging on the wires connecting the circuit board to the AAA cell holder or the F°/C° switch on the back.

Step 2

Release the circuit board. With a miniature Phillips screwdriver, remove the two retaining screws from either end of the circuit board. Carefully flip the board and LCD to view the front, component side of the board and LCD.

Layout

As shown in Figure 8-2b, the back of the circuit board is sparsely populated. Other than the connections to the F°/C° switch and the AAA cell, the only significant component on the board is the white, rectangular humidity sensor.

The front of the board, shown in Figure 8-3, holds an epoxy dipped thermistor, a cylindrical crystal oscillator, a few SMT resistors and capacitors, and the two momentary contact switches for max/min and reset functions. The microcontroller, hidden under the black epoxy blob, and the adhesive ribbon cable taped to the LCD are also on the front of the circuit board.

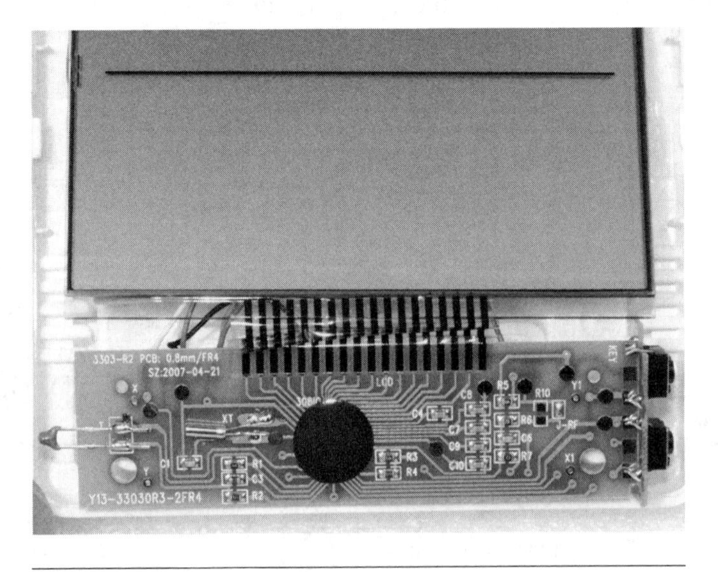

FIGURE 8-3 Front circuit board layout

Components

The microcontroller may be the most advanced component on the circuit board, but the most relevant for this teardown are the temperature and humidity sensors.

Thermistor

The epoxy-dipped thermistor, or temperature-dependent resistor, shown in Figure 8-4, is mounted so that its body is suspended near the air ducts on the plastic case and away from the heat-producing components on the circuit board. This is a negative temperature coefficient (NTC) thermistor, meaning that as temperature increases, the resistance decreases. For example, I measured the resistance of the thermistor in my meter as 8.6KΩ at 80°F and 10KΩ at 60°F.

The ideal thermistor has constant relative changes in resistance for given changes in temperature, regardless of the absolute temperature. In other words, if you plot temperature versus resistance, the relationship is a straight line. In practice, inexpensive thermistors exhibit a nonlinear relationship between temperature and resistance. This nonlinearity can be corrected by placing a resistor with a complementary temperature coefficient in parallel with the thermistor. A second approach is to program a microcontroller either to use a lookup table or execute a corrective formula to linearize the response.

FIGURE 8-4 Thermistor (T) and crystal oscillator (XT)

Humidity Sensor

The relative humidity (RH) sensor used in this meter is a resistive sensor, shown in Figure 8-5, which consists of a moisture-absorbing substrate and conductive traces. With increased moisture in the air, more moisture enters the substrate and the resistance drops. In addition, the moisture in the substrate alters the capacitance between the conductive traces.

The sensor's *impedance*—resistance of the substrate and capacitive reactance of the sensor—is inversely proportional to relative humidity. Because the relationship between impedance and relative humidity is roughly exponential, the raw reading from the sensor must be linearized with a lookup table or formula. The impedance of the RH sensor in my meter measured 400KΩ at 36 percent RH and 40KΩ at 98 percent RH.

The more expensive and more accurate digital hygrometers use capacitive relative humidity sensors that consist of an exposed dielectric (insulating material) between two conductors. Increases in relative humidity increase the dielectric constant of the insulator. Unlike resistive sensors, this response is linear over a wide range of temperatures and relative humidity conditions.

A third type of relative humidity sensor used with digital hygrometers measures the thermal conductivity of the air relative to a sample of dry air. In general, thermal conductivity sensors are used in high-temperature applications where capacitive and resistive sensors would not survive.

A fourth type of relative humidity sensor consists of a quartz crystal with a hygroscopic coating. An increase in relative humidity increases the dimensions of the quartz crystal and causes a downward shift in the resonant frequency. Definitely overkill for a home comfort gauge.

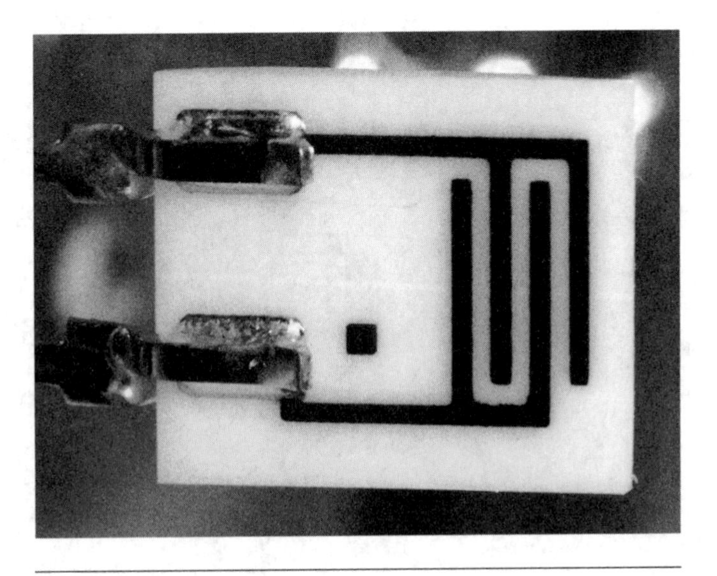

FIGURE 8-5 Humidity sensor

Microcontroller

The microcontroller hidden under the glob of black epoxy handles the LCD display and the momentary switches, and reads signals from the sensors. (See the discussion of how to handle these epoxy blobs in Chapter 3.) For our purposes, what's important is that the microcontroller reads the two sensors, processes the data, and displays the results.

Crystal Oscillator

The cylindrical tuning fork crystal oscillator, which operates at 32.768kHz, can be seen on the circuit board next to the thermistor in Figure 8-4. This is a popular, inexpensive, low-power clock source for microcontrollers and microprocessors, in part because the operating frequency is stable over a relatively wide temperature range.

Many microcontrollers use an onboard crystal oscillator or an external resistance-capacitance (RC) controlled oscillator. However, in applications for which timing is critical, an external, high-accuracy crystal oscillator is often best. As you'll soon see, this is one of those applications.

LCD Panel

The LCD panel, the largest component in this meter, is notable in the edge connector used to mate it with the microcontroller and circuit board. As shown in Figure 8-3, the transparent ribbon cable connects pads on the circuit board to those on the edge of the LCD. The pressure connection is secured by a piece of clear tape over each end of the cable.

One of the limitations of LCD displays over competing display technologies, such as LED displays, is operating temperature range. For example, I run with an LCD watch. When the ambient temperature dips below freezing, the digits on the LCD fade to gray. I haven't verified it, but I assume that given the manufacturer lists the accuracy down to 14°F (–10°C), the LCD is capable of operating at that temperature.

Some manufacturers avoid the limitations imposed by the LCD operating environment by adding remote sensing capabilities to their hygro thermometers. This is frequently the design of weather monitors in which the display is kept in a warm room and the sensor is placed outside.

How It Works

The mechanical, analog technologies for measuring temperature and relative humidity—from bimetallic and mercury thermometers, to hygrometers made from strands of horse hair or a sheet of animal intestine—have stiff competition from electronic meters. While mercury thermometers are typically more accurate than similarly priced digital thermometers, there is the issue of environmental toxicity from the mercury. Instruments that rely on the elongation and shortening of horse hair or the tension on a piece of animal intestine stretched over a frame as a function of relative humidity require periodic calibration.

Digital thermometers with resistive sensors rely on the fundamental physical property that the electrical resistance of a conductor varies with temperature. An ordinary metal film resistor, such as a length of copper wire, displays changes in resistance as a function of temperature. A thermistor simply displays this property with a relatively small change in temperature. As noted earlier, the thermistor in this meter is designed to offer decreased resistance as the temperature increases.

You can use several methods to determine the resistance of a thermistor in a digital thermometer. One method is to apply a fixed voltage across the thermistor and measure the current. Another less hardware-intensive approach is to use the thermistor as the resistance component of an RC (resistor-capacitor) circuit and to measure the time required for an applied voltage to drop to a specific value. Recall that the time constant of an RC circuit is the time required for the voltage to drop to 37 percent of its initial value. Mathematically, this is expressed as follows:

$$T = RC$$

where T = time in seconds, R = resistance in ohms, and C = capacitance in farads.

Based on my measurements and assessment of the circuit, our digital thermometer uses the later approach. Every 5 seconds, a series of a dozen pulses, 4V peak-to-peak at 12.5kHz, is applied to the thermistor in series with SMT capacitor C1. The greater the temperature, the lower the resistance of the thermistor and the shorter the time constant. In other words, the greater the temperature, the faster the decay in voltage across C1.

Note the three I/O leads from the microcontroller that interface with the thermistor: R1, R2, and C3 in Figure 8-3. This is reflected in the simplified schematic, shown in Figure 8-6.

For simplicity, assume that the microcontroller reads an input pin as high if the voltage is at or above 0.63V and low for anything below 0.63V. With an accurate clock on board, the microprocessor times how long it takes for the voltage to drop so that the input pin transitions from high to low. This is repeated for each of the pulses, and the average value is used to determine the resistance of the thermistor. Given the resistance, the microcontroller uses a lookup table or a mathematical calculation to convert the thermistor resistance to a temperature value that is then displayed on the LCD.

Note that there are two connections to the C3-thermistor junction. Given that only one connection is required to monitor the voltage across C3, the second lead is probably for auto-calibration or a self-test.

Now let's focus on the hygrometer. You may have noticed that it's difficult to find a digital hygrometer without a thermometer. This is because an accurate calculation

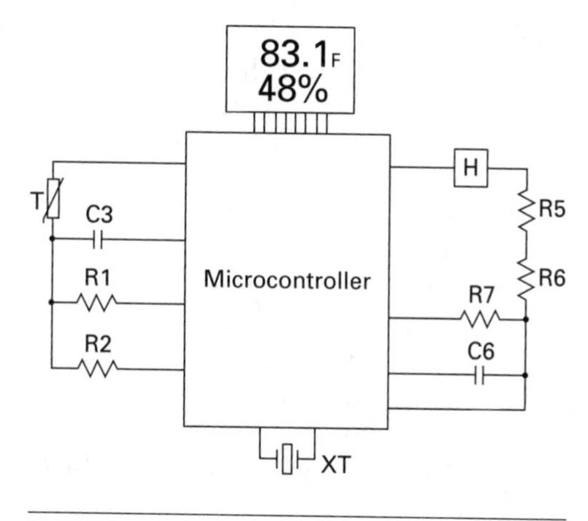

FIGURE 8-6 Simplified schematic

of relative humidity requires a value for the ambient temperature. Because the ambient temperature is used to compute relative humidity and the display circuitry is already in place, it makes sense to add value to the meter by displaying the temperature.

The hygrometer sensor wiring to the microcontroller is similar to that used for the thermistor, as shown in Figure 8-6. One terminal of the humidity sensor is connected directly to an I/O pin of the microcontroller. The other terminal is connected in series with two resistors, R5 and R6, and then to three I/O pins via C6, R7, and a direct connection to the microcontroller. As with the thermistor, every 5 seconds, a series of square waves, in this case about 3VPP (peak-to-peak) at 12.5kHz, is applied to the humidity sensor. The time constant is read by the microcontroller and, along with ambient temperature, is used to calculate and display relative humidity.

Mods

As discussed in Chapter 7, you know that one use for a digital hygro thermometer is to function as part of a closed loop system that controls a heating, cooling, and-or humidification system. The challenge is getting at and interpreting the output of the microcontroller.

Adding Calibration Capabilities

One of the limitations of this meter is that it offers no way of easily calibrating the temperature or relative humidity. You're either stuck with whatever is displayed on the LCD or you can apply correction factors to the values on the display. You can add some adjustability by inserting a subminiature potentiometer in series with the thermistor. If you need to lower the resistance presented by the thermistor, try the potentiometer in parallel with the thermistor. In either case, make certain you use a potentiometer with a low temperature coefficient.

You can try the same approach for the relative humidity sensor. Do your best to avoid handling the sensor, because oils from your fingers can permanently alter the resistance of the sensor.

An obvious issue in calibrating your meter is that you need a standard. Temperature is usually less of a challenge, because accurate analog thermometers are so readily available. Relative humidity is another issue. If you can, borrow a 2 percent hygrometer designed for meteorology to use as a standard. Another option is to bring your meter to a store with high-end hygrometers on display and compare settings after you've given your meter time to stabilize.

Chapter 9

Stereo Power Amplifier

Whether your passion is 7.1 channel surround sound for video games and home theatre or playing vinyl records in stereo, abundant, clean, affordable audio power is a must. And, thanks to specialized, high-performance audio amplifier ICs, audiophile-quality amplifiers are commodity items. All you need to build a state-of-the-art amplifier are a pair of amplifier ICs, a handful of discrete components, a robust DC power supply, and a heat sink large enough to keep the ICs from going into thermal shutdown. The usual amplifier design headaches, from impedance matching and short protection, to providing a standby mode, are handled by the ICs.

In this chapter, I'll tear down the AudioSource AMP 100, shown with its cover removed in Figure 9-1. This Chinese-manufactured stereo amplifier, which retails for about $140, is popular in the audio community because of its simplicity, specifications, and affordability. As a teardown, it's a great introduction to integrated amplifier chips, operational amplifiers, amplifier specifications, and concepts such as bridging. At the end of the chapter, I'll offer a mod that addresses a minor complaint that some have with the amplifier.

Highlights

At first, you might find it intimidating to tear down something as seemingly complex as a stereo amplifier. However, thanks to the simplicity afforded by the power amplifier ICs in the AMP 100, you won't be dealing with dozens of transistors and hundreds of support components. If you can handle a screwdriver, you can handle this teardown. As you'll see, most of the circuit analysis revolves around two circuit boards: the power board containing the power supply and power amplifier ICs and the smaller but more complex logic board.

During the teardown, note the following:

- The simplicity of the power board
- The size, mass, and orientation of the heat sink

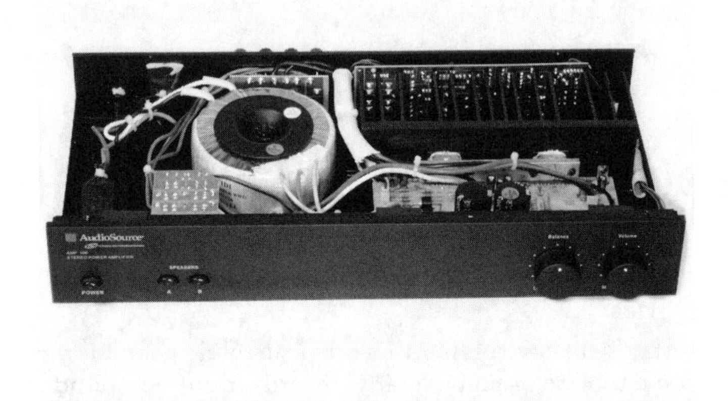

FIGURE 9-1 AudioSource AMP 100 stereo amplifier

- The mass of the toroidal power transformer
- The pair of zener diode voltage regulators

Specifications

The specifications of the AMP 100 reflect those of the two integrated amplifier ICs. Highlights of the specifications, extracted from the user manual and AudioSource web site (www.rodinaudio.com), include the following:

- Frequency response: 20Hz to 20kHz ± 1dB; 5Hz to 65kHz ± 3dB
- Signal-to-noise ratio (SNR): 101dB
- Maximum stereo power output: 60W at 8Ω, 1 percent total harmonic distortion plus noise (THD+N)
- Maximum bridged-mono output: 160W at 8Ω, 1 percent THD+N
- Power consumption: 500W
- A, B, A+B speaker selector switches
- Auto-on (signal sensing)
- Auto input switching
- Front panel controls: power, speakers, volume, and balance
- 3 × 16 1/2 × 9 1/4 inches (HWD)
- 9.6 pounds

Additional specifications, based on the datasheet for the power amplifier chips, include Class AB operation, auto-thermal shutdown, and short-circuit protection.

Together, the specifications detail a tough, compact amplifier. But what do the individual specifications mean? I'll consider the more significant specifications one at a time.

Frequency Response

An often-touted audio amplifier specification is the frequency response or bandwidth, typically at the 3 and 1dB points. The ideal frequency response for a home entertainment amplifier is *flat*—that is, ±0dB through audible audio spectrum. In the context of amplifier bandwidth, a dB, or decibel, is a logarithmic relationship between the power output measured at two frequencies within the bandwidth, defined by this equation:

$$dB = 10 \log_{10} (P1/P2)$$

where P1 and P2 are power measures taken within the specified bandwidth. For example, let's say a 4kHz tone is fed to the amplifier input and the output power is measured as 50W. A tone of the same amplitude but frequency of 20Hz is fed to the amplifier input and the output is measured as 45W. Is the amplifier working within spec? Let's apply the formula:

$$dB = 10 \log_{10} (45/50)$$

$$dB = -0.45$$

The output is within the ±1dB limit, and therefore in spec. In other words, according to the manufacturer, the power output doesn't fluctuate more than 1dB above or below the nominal power output over the frequency range of 20Hz to 20kHz.

The 1dB point is relevant because it's roughly equal to the just noticeable difference. That is, while most people can't distinguish between 45 and 50W (0.45dB), most can differentiate between 40 and 50W (1dB) and certainly between 50 and 25W (3dB).

The 3dB point is relevant because it represents the half-power point, or where the amplifier output drops to half of the nominal output. In the case of the AMP 100, power drops to half at 5 and 65Hz. For the price of this amplifier, both the 1 and 3dB bandwidth figures are excellent.

SNR

The signal-to-noise ratio is a measure of the amplitude of the music or other desired signal to the amplitude of the background noise. SNR is also measured in dB:

$$SNR (dB) = 10 \log_{10} (PSignal/PNoise)$$

where PSignal is the power of the desired signal and PNoise is the amplitude of the noise. From a practical perspective, noise is what you hear when the volume is turned up but no external signal is applied to the amplifier, or when there's a pause between songs. Ideally, SNR is infinite. However, noise is inescapable above absolute zero. For this reason, passing a signal through a resistor introduces noise into the signal and diminishes the SNR.

Considering the price of the AMP 100, an SNR of 101dB is good, if you make a number of assumptions. For example, I picked up an old Marantz integrated amplifier on eBay that was originally an order of magnitude more expensive than the AMP 100. It has an SNR of 90dB, and I'm certain it outperforms the AMP 100. As with power, you can't simply compare two SNR figures without specifying the bandwidth and frequencies at which the signal and noise are measured.

According to the AudioSource web site, the SNR of the AMP 100 is 101dBA, referred to 50W, at 8Ω. The dBA or A-weighted dB is commonly used in measuring the noise of audio equipment, but the reference to 50W at 8Ω is incomplete. A more informative description would include assurance of CEA-490-A R-2008 compliance. This standard of the Consumer Electronics Association (www.ce.org), which defines test conditions and test measurement procedures for determining various performance characteristics of home audio equipment, enables consumers to compare audio amplifiers on an equal basis. A standards organization sometimes cited by manufacturers as a reference for SNR is the Institute of High Fidelity (IHF), now part of the Electronic Industries Alliance (www.eia.org). Unfortunately, most audio amplifier manufacturers simply offer an SNR in dB, with no further elaboration.

THD+N

Total harmonic distortion plus noise is a way of quantifying audio "crud." As you crank up the volume of this or any other home entertainment audio amp, the sound becomes increasingly distorted, noisy, and rich in harmonics. In the context of the AMP 100, harmonics are undesirable. However, this doesn't hold for special-purpose amps, such as guitar amps, which are designed to distort sounds. (See Chapter 14 for more details.)

You can probably get an intuitive handle on THD+N by studying the following formula:

$$THD + N(bw) = (\text{sum of harmonic powers} + \text{noise power})/\text{output power}$$

The first thing to note is that THD+N makes sense only when bandwidth (bw) is specified, typically the 1 and 3dB bandwidths. The sum of harmonic powers is the sum of the powers of the harmonics: for example, the power of the second harmonic plus the power of the third harmonic, and so on. Noise power is the power of crud produced by the audio system without an input signal, due to, for example, random movement of electrons within conductors, interference from other devices, and component imperfections. Output power is the power output of the fundamental frequency when the THD measurement is taken.

Common fundamental frequencies for evaluating THD+N are 1 and 4kHz. Increasing the output power typically increases the sum of harmonic powers. For example, if the power is increased at 4kHz, you'd expect the powers at 8, 12, 16, 20kHz, and so on, to increase as well.

THD+N is a ratio measurement expressed as a percentage. Ideally, THD+N is somewhere near 0 percent, but 1 percent with the volume cranked up isn't terrible. One way to reduce THD+N is to reduce output power to well below maximum output power. If a 100W amplifier is loafing along at, say, 50W, then the harmonic power will be down because of the increased dynamic range or headroom available. Background noise isn't as responsive to decreased output power as THD.

Power Output

As the specifications suggest, simply stating that an amplifier has 100W output is incomplete and misleading. The best way to interpret the power output rating of the AMP 100 is that, given an 8Ω load per channel and the limitation of 1 percent THD+N, it will deliver up to 60W per channel, or 120W total. The specification doesn't say 60W is the maximum output power—you may be able to get 75W from the unit, but under different assumptions. For example, if you assume 4Ω speakers and a limitation of 1 percent THD+N, the maximum output power is 75W. Similarly, if you assume 8Ω speakers and up to 10 percent THD+N, you might be able to extract up to 80W from the amplifier. The point is, when you are evaluating amplifier output power and other specifications, make certain you understand the operating assumptions.

A feature of this amplifier that makes it appealing to home theater enthusiasts is that the output is *bridgeable*, meaning both channels of the amplifier can be used to drive a single speaker. Let's say you want to use the AMP 100 as the center (mono) channel of your home theater setup. As with any other stereo amplifier, you could feed the same signal to both left and right input channels of the amplifier, and the left and right speakers would reproduce the same in-phase output.

However, because this is a bridgeable amplifier, you have the option of connecting a single speaker between the left and right output terminals. In the bridge mode, the AMP 100 is configured so that the left and right channels are 180 degrees out of phase. As a result, if each channel swings from 0 to 20V relative to ground, the bridged output will swing 0 to 40V relative to each output terminal. For example, at the instant when the right channel terminal is +20V relative to ground, the left channel terminal is –20V relative to ground, a 40V differential. Otherwise, if the output signals were in phase, which is the case in a regular stereo setup, there would be no voltage difference between the output terminals.

Bridging provides greater output power than simply sending the same signal to each channel of a stereo amplifier. For example, the AMP 100 produces an additional 40W as a bridged-mono output amplifier compared with the stereo out, assuming the same speaker impedance and THD+N. To understand how this is possible, recall the power equation

$$P = V^2/R$$

where P is power in watts, V is the voltage in volts, and R is the resistance in ohms. Let's assume that each channel of a stereo amplifier produces a maximum of 20V across an 8Ω speaker. That's $20^2/8$, or 50W, per channel for a total of 100W. Now consider the bridged output with the same output and speaker impedance. The voltage maximum is now 40V, producing $40^2/8$, or 200W. Although the AMP 100 doesn't provide the theoretical 100W of additional power in bridge configuration, the additional 40W is an increase of just more than 1dB, a noticeable increase in output.

Power Consumption

The AMP 100 is rated at 500W power consumption, and much of this power is destined for the heat sink, as opposed to the speakers. The power consumption, which reflects the efficiency of the amplifier, is largely due to the power requirements of the power ICs, which operate as Class AB amplifiers. Recall that an amplifier configured for Class A operation conducts through the entire input cycle. Using an oscilloscope, you'd see sine wave input and sine wave output. A transistor, IC, or tube amplifier configured for Class B operation, in contrast, conducts for half of the input cycle. On your scope, you'd see sine wave in either the positive or negative half of the sine wave out, but not both.

A pair of Class B amplifiers, configured in a push-pull arrangement, with one assigned to the positive and one to the negative component of the input signal, can faithfully amplify the full sine wave. The advantage of Class B over Class A operation is increased efficiency. The price is increased THD, component count, and, in general, complexity.

Class AB operation is akin to Class B operation with a little Class A operation thrown in to decrease the THD. The Class AB amplifier ICs conduct for a little more than half of each input cycle. The result is significantly improved linearity and reduced THD compared with a Class B configuration, because the crossover from one amplifier to the other is smoother. The price for reduced THD is decreased efficiency.

Auto-On

A feature of this amplifier that you'll either love or hate is the auto-on feature, which is essentially active standby. With the power button depressed, the amplifier is on standby until it detects an audio signal on one of its input connectors. Auto-on is trivial to implement because of a pin dedicated to this function on each power IC.

The advantage of auto-on is that it's one less button to press if you're booting up your home entertainment system, assuming you don't have a power strip with an on–off switch. One disadvantage, in this era of carbon credits and green thinking, is that this is just one more device to add to the dozens of devices consuming energy in your home. Another disadvantage is that you're forced to use this feature.

If the AMP 100 is on and there's a long pause in a music passage, the unit responds by shutting down until triggered by another signal. In this condition, you'll hear the relay click open and closed and the audio will stutter. I haven't experienced this problem, perhaps because I don't watch movies with lots of silence in the soundtracks, but it's a hot topic on the forums.

Auto-Input

Auto-input is like auto-on, in that the AMP 100 automatically senses an audio input signal and then configures itself accordingly. In this case, however, the configuration has to do with selecting the signal source. The front panel is void of a source selector switch. As a result, you have plug in the output cables from two audio devices when you initially configure the system. The amplifier will automatically switch to the active input, line 1 or line 2, when there is signal present. If a signal is present on both input channels, line 1 takes priority.

Form Factor

The specifications relating to physical size and mass are straightforward. Of note is that much of the bulk and mass of the AMP 100 are due to the power transformer and heat sink.

Operation

This amp is plug-and-play. The only significant operating decisions are at setup, when you use the two slider switches in the rear of the unit to select between either standard stereo output or bridged mono output and between normal on–off and auto-detect. The balance and volume controls function as you'd expect.

Teardown

You should be able to complete the teardown, illustrated in Figure 9-2, in about an hour.

Tools and Instruments

You'll need a few Phillips-head screwdrivers to disassemble the housing and extract the circuit boards, and a 10mm nut driver to remove the two potentiometers from the front panel. A multimeter will be useful for tracing the circuits. If you want to trace signals through the circuit boards, a dual-trace oscilloscope is a real time-saver. You might also find a muffin tin useful to store and sort the various screws produced by the teardown.

FIGURE 9-2 Teardown sequence

FIGURE 9-2 ***(continued)*** Teardown sequence

Step by Step

I found it easier to examine the live circuitry by extracting the transformer from the chassis, which involves unsoldering the leads to the fuse and power cord. You'll have to connect another power cord to the transformer, preferably through a Variac or other fused source, to work with the two circuit boards. If you decide to extract the transformer, you'll need a soldering iron, a crescent wrench to remove the toroidal transformer, and another ten minutes for the teardown.

Step 1

Remove the top cover. Remove the machine screws that secure the top cover. You'll find two on the top, near the faceplate; two screws on either side of the unit; and three in the back. Lift the cover up and back and set it aside. Figure 9-2a shows the view from the rear of the unit. Note the relative positions of the input and output jacks and the mode switches on the rear panel, the large heat sink on the left, and the massive toroidal transformer on the right.

Step 2

Remove the faceplate. The AMP 100 has a cosmetic faceplate attached to the chassis. Figure 9-2b shows the faceplate and chassis adjacent to the power board. Note the potentiometer shafts running through the chassis at the upper-left of the figure.

To remove the faceplate, first pull off the volume and balance knobs from the potentiometer shafts using your hands. You shouldn't require tools, but if you do, protect the faceplate from nicks and scratches with a cloth or thin piece of plastic or wood. Next, remove the screw on top and two screws on the bottom of the faceplate. Set the faceplate aside with the cover.

Step 3

Release the front panel components. Remove the 10mm nuts from the shafts of the volume and balance potentiometers. Because the front of the chassis is hidden by the faceplate, you can use a crescent wrench or other means to remove the 10mm nuts, even if you scratch the chassis. However, if you have a hollow shaft nut driver, you should be able to remove the nuts without leaving a mark.

Remove the four screws securing the power and speaker select push buttons to the front of the chassis, as shown in Figure 9-2c. The speaker select push button assembly is shown in Figure 9-2d. The larger wires carry the audio output while the two pairs of smaller wires supply the LEDs.

Step 4

Release the rear panel components. Rotate the unit so that you're facing the rear of the chassis, and remove the screws securing the speaker output lugs on the left. As shown in Figure 9-2e, the left and right channel connections on the speaker output board are electrically isolated from each other. Remove the five machine screws securing the logic board to the right rear of the chassis. This board contains the RCA connectors.

Step 5

Extract the large circuit board and heat sink. Next, remove the screw securing the large power board to the chassis. Note that the heat sink is attached to power amplifier ICs on the power board, as shown in Figure 9-2f. Now, from the bottom of the amplifier, remove the three screws securing the heat sink. Figure 9-2g shows the extracted power board and heat sink assembly from the perspective of the circuit board bottom. The shafts of the volume and balance potentiometers are at the upper-right of the figure.

Step 6

Release the power cables. Disconnect the three-wire transformer secondary cable connector and the two-wire power cable connectors leading to the power board. Figure 9-2h shows the cable connectors on the power board. The two-wire connectors at the bottom of the figure feed the LEDs within the power and speaker selection push buttons. Set aside the main chassis with the toroidal power transformer, power cord, and fuse holder.

Step 7

Release the power ICs from the heat sink. This will allow you to get a good look at the ICs. If you're following along with your amplifier at home, note that the ICs are electrically insulated from the heat sink. If you reassemble the amplifier without ensuring that the insulation is intact, you'll fry the ICs and likely damage the power supply.

Remove the four machine screws securing the two power ICs to the heat sink. The screws run from the heat sink to an aluminum bar that secures the two power ICs to the heat sink. Remove a fifth screw that secures the power board to the heat sink. Note the mica insulators and conductive grease between the heat sink and the metal tabs on the power ICs. Keep the two mica insulators and conductive grease with the heat sink and move the heat sink to a safe place where the mica won't be disturbed. Figure 9-2i shows one of the power ICs unattached to the heat sink.

Step 8

Extract the toroidal transformer. If you want to test the live circuit, you should extract the toroidal transformer from the chassis and supply the primary with 120VAC. Otherwise, you'll be constrained by the bulk of the chassis. First, unsolder the primary winding connections to the line cord and fuse. Next, remove the bolt running through the center of the toroidal transformer, shown in Figure 9-2j, and through the chassis bottom. Remove the transformer and move the chassis to the side. Solder a new power cord to the transformer primary, using electrical tape or shrink-wrap tubing to insulate the solder connections. Don't forget to use a fused source when you plug in the cord.

Layout

The internal layout of the AMP 100 is worth studying. There's no wasted space, and yet the airflow through the heat sink isn't compromised. During the teardown, we encountered seven subcomponents:

- Toroidal transformer, mounted directly to the chassis. The primary is soldered to the fuse and power cord, while the secondary terminates in a three-conductor connector that mates with the power board near the four silicon rectifiers, as shown in Figure 9-2h.
- Power board that contains the power supply, power amplifier ICs, an operational amplifier, and the volume and balance potentiometers. The board is secured by the potentiometer shafts to the front of the chassis, by the aluminum heat sink, and by a metal standoff connected to the chassis bottom.

- Large anodized aluminum heat sink connected to the two power ICs. The heat sink is positioned near the center of the chassis, immediately above the ventilation slots in the base of the chassis and below the ventilation slots in the cover. See Figure 9-2f.
- Logic board with two pairs of RCA audio input jacks for Line 1 and Line 2, one pair of output jacks, mono/stereo and normal/sensing switches, four operational amplifiers, a transistor, and a relay.
- AC power front panel switch assembly. Embedded within each switch is an LED that indicates a depressed condition. The power board supplies power for the LEDs.
- A/B signal selection front panel switch assembly. This switch assembly accepts the right and left channel outputs from the power board and distributes the output to the appropriate terminals on the audio output connector assembly. As with the power button, an LED within each button receives power from the power board. See Figure 9-2d.
- Audio output connector assembly. The output connector, which accepts banana jacks as well as tinned wire, is a little too cramped for my tastes. Fortunately, connecting speakers to the amplifier is usually a one-time affair.

To get the most out of this exercise, disconnect the push button and audio output assemblies from the two circuit boards, as shown in Figure 9-3. This will enable you to examine the layout of the two boards in more detail, with unrestricted access to components while probing for signals.

You can avoid the hassle of connecting the output to speakers by terminating the left and right audio circuits with 8Ω power resistors. If you keep the volume

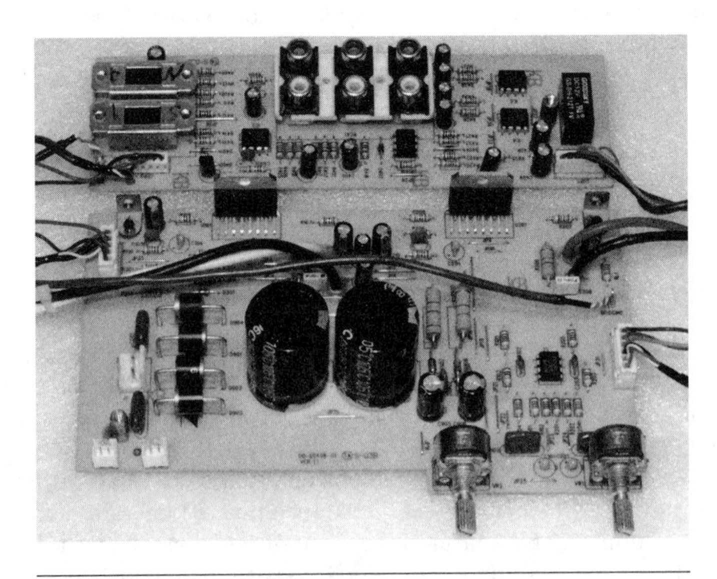

FIGURE 9-3 Circuit analysis setup

potentiometer turned down, you can get away with a 10W resistor on each channel. In addition, because the ICs aren't connected to a heat sink, limit power-on times to a few seconds. Otherwise, there's a risk of thermal shutdown, and you'll have to wait a few minutes for the integrated amplifier ICs to cool before you can return to testing.

Now let's focus on the larger power board and its connections. Orient the board so that the two potentiometers are on the right, as shown in Figure 9-3. In this position, the TDA7294 IC that powers the left channel is on the left and the TDA7294 for the right channel is on the right. The audio output from each power IC exits the board immediately to the right of each chip. Three cables connect the two boards: a six-conductor cable on the right, a five-conductor connector on the left, and a separate bridging cable that extends from the lower-left edge of the logic board to the right edge of the power board, near the right audio output connection.

The five- and six-conductor cables each include a miniature coaxial cable with dual center conductors for signal right and left. The coaxial shield is signal ground. The remaining wires in the six-conductor cable are ±15VDC and power supply ground. The five-conductor cable carries 15VDC and power supply ground.

The bridging cable is worth examining closely. It consists of a coaxial cable containing a twisted pair of wires. The shield is not terminated on either end of the cable, and both inner wires are connected together to solder posts on the circuit boards. Why not use a single wire for the inner conductor? Why use coaxial cable?

The power supply components, less the toroidal transformer, are located on the front half of the power board. Starting on the left edge of the board, we have a three-pin connector for the secondary of the toroidal power transformer, four silicon power diodes, Mylar bypass capacitors, and two 6800µf at 50VDC electrolytic capacitors. Note how the power diodes are mounted above the board for heat dissipation. Immediately behind the balance potentiometer are a pair of zener diodes and fusible resistors, as shown in Figure 9-4. The large filter capacitors feed the power ICs directly with ±36VDC, while the zener diodes provide ±15VDC voltage for the other ICs, a transistor, and a relay.

This is a good point at which to note that 36VDC is enough to cause you serious harm. Moreover, because this is a dual-voltage supply, you could find yourself in contact with 72VDC if you're not careful. If you decide to work with the live circuit, don't become complacent simply because you're dealing with solid-state components.

Behind the volume control is a 5532D dual low-noise operational amplifier, as shown in Figure 9-5. It receives audio input from the adjacent cable and feeds the inputs of the two TDA7294 ICs. Other components on the power board worth noting are the two dozen, five-band, 1/8W metal film resistors; four polarized tantalum capacitors associated with the power IC; and each potentiometer. Most of these components are visible in Figure 9-5. There are no ferrite beads, chokes, or other inductors on the board.

Figure 9-5 also shows the main board with the connection to the heat sink intact. From this perspective, you can see the aluminum compression bar and the thermal grease between the amplifier ICs and the heat sink.

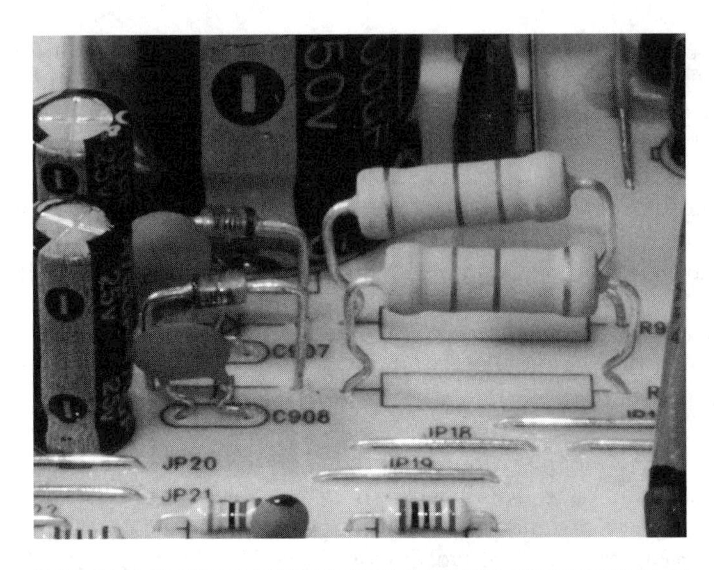

FIGURE 9-4 Zener diodes and fusible resistors on the power board

FIGURE 9-5 Right side of power board

One of the features of the single-sided power board is that components are labeled in a logical manner. Components associated with the right channel are numbered in the 200s, such as U201, R205, and C204. Components associated with the left channel are numbered in the 100s, and the labels are applied symmetrically. For example, U101 and U201 are both power ICs and

C104 and C204 are tantalum capacitors of equal value used with U101 and U201, respectively. This labeling makes signal tracing and creating a schematic a breeze.

The logic board is more compact, more complex, and more of a challenge to sort out, compared to the larger power board. Starting on the left of the board, using Figure 9-6 as a reference, you see two DPDT (double-pole, double-throw) slider switches. The top switch is for the switching/normal setting and the bottom switch is for setting the mono/stereo preference. The switches are designed to prevent accidental changes in the settings. There's a slot to use a coin or screwdriver blade to slide each switch. Continuing to the right of the switches is the only transistor on the board, a C2235 NPN transistor in a TO-92 package. To the right of the transistor is a 4558D dual general-purpose operational amplifier in an 8-pin DIP (dual inline package).

Now, moving just to the right of center, in front of the two RCA signal output jacks, is a second 4558D, as shown in Figure 9-7. To the right of the RCA connectors are a pair of 5332D operational amplifiers, identical to the one used on the power board. Finally, there's a 12VDC DPDT relay on the right edge of the board, immediately behind the 6-pin connector. As shown in Figures 9-6 and 9-7, there are a dozen small electrolytic capacitors, a few dozen five-band metal film resistors, and a few signal diodes on the board.

Like the power board, the single-sided logic board is labeled, but other than the ICs, switches, and relay, components are numbered in the 400s. To add to the modest challenge of tracing the circuitry on the logic board, there are jumpers and resistors under the RCA connector block. Don't overlook these as you follow the signal path through the circuitry on the board.

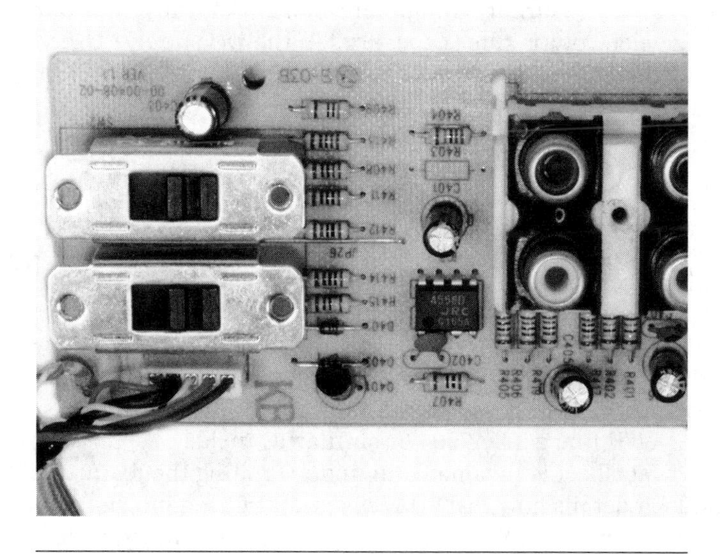

FIGURE 9-6 Left side of logic board

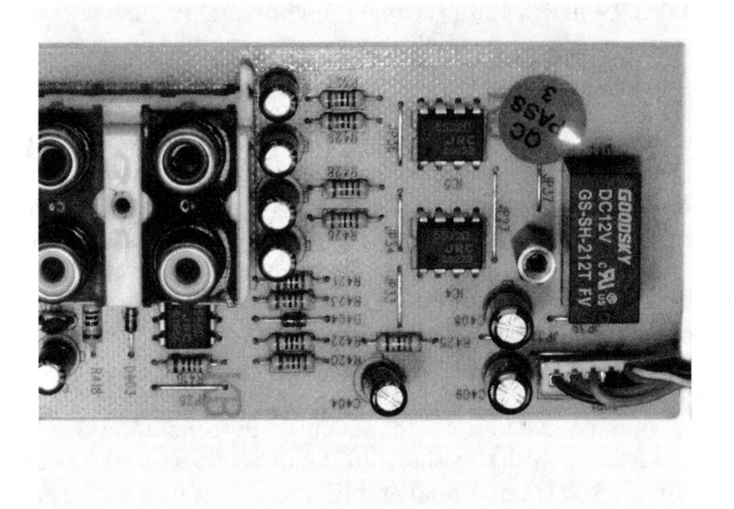

FIGURE 9-7 Right side of logic board

Components

If you're following along with an AMP 100 on your bench, take the time to examine each component carefully as it's discussed. Although I wouldn't unsolder the ICs unless you plan to upgrade the amplifier with higher performance chips, anything else is fair game. Not sure of the value of one of those five-band resistors? Unsolder one end and measure it with an ohmmeter. Want to check out the adequacy of the large electrolytic filter capacitors in the DC supply? Use your oscilloscope to monitor the DC voltage while you drive the amplifier at full output. Just exercise the usual caution around live circuits.

TDA7294 Power ICs

It should be no surprise by now that the two TDA7294 power ICs define the capabilities of the AMP 100. According to the datasheet from STMicroelectronics (www.st.com), each 15-pin IC features a MOS power amplifier that can deliver output current of ±10A when supplied with 80VDC. The IC is surprisingly affordable; at the time of writing, a single unit price was less than $6 from Digi-Key (www.digikey.com). If you own a subwoofer or self-powered speaker, there's a good chance you'll find a TDA7294 or similar IC inside.

The datasheet is invaluable in understanding the features of the TDA7294. For example, it details how the TDA7294 is configured internally to operate with another TDA7294 in bridge mode. What's surprising is the marked similarity between the suggested circuits in the datasheet and the AMP 100 circuitry. By the way, if you're trying to figure out the pin numbering on this oddly packaged chip, the first pin is on the far left if you're facing the front of the package, as shown in Figure 9-2i.

The TDA7294 is theoretically capable of delivering 200W output power, operating as a Class AB amplifier. This level of power is theoretical because it assumes a high-efficiency heat-dissipation system that is practically unachievable. As it is, the heat sink in the AMP 100 occupies a greater volume than that of the two circuit boards combined.

Fortunately, the TDA7294 is internally protected against thermal meltdown. It automatically switches to a muted state when its internal temperature reaches 145°C (293°F) and to standby at 150°C (302°F). In addition, even if you use liquid nitrogen to cool the ICs, the THD without noise exceeds 10 percent when the chip is pushed above 85W output.

Muting and standby functions can be controlled externally as well, independent of the thermal protection circuitry. Muting the audio output of the power chips is useful at startup to avoid switching noise as the AMP 100 selects the input channel and activates the auto-on function. Standby is a power-saving feature that supports the auto-on function as well. Without standby, the ICs and heat sink would be required to dissipate heat at all times.

In addition to mute and standby functions, automatic thermal shutdown, and short-circuit protection, each IC maximizes efficiency by switching on the internal power transistors only during the portion of the signal when more output voltage is needed. Although the IC has separate voltage supply pins for this power transistor circuit and the remainder of the chip, they are tied together in the AMP 100.

The TDA7294 requires only a handful of discrete capacitors and resistors to define its input impedance, gain, standby time, and mute time, and to bypass supply voltage inputs. In addition, the IC doesn't require a highly regulated DC power supply because its supply voltage rejection (SVR) is at least 60dB. SVR, also referred to as Power Supply Rejection Ratio, or PSRR, is a measure of the change in supply voltage divided by the simultaneous change in output voltage, if any. The ideal SVR is infinity, meaning that the supply voltage can swing wildly, while the output voltage remains unaffected. Even so, an SVR of 60dB allows for a simple power supply design.

The metal tab on the IC is connected to the negative voltage supply pin, which is at –36VDC on the AMP 100. Not only does the exposed tab present a shock hazard if you're working with a hot circuit, but if you reattach the heat sink without properly insulating the tab, you'll short-circuit the power supply through the IC.

Extruded Aluminum Heat Sink

As noted in the discussion of the power amplifier ICs, the limiting factor in extracting power from the TDA7294 is removing heat. If you recall from basic physics, heat is transferred by conduction, radiation, and convection, from an object that's at a higher temperature than its surroundings. The AMP 100 relies on all three methods of heat transfer.

The power ICs are physically connected to a large heat sink that is cooled by natural convection. The chassis ventilation is designed with slots immediately below and above the cooling fins, maximizing the flow of heat carrying air, or convection currents, from below to above the unit. Heat from the ICs is conducted through the metal tabs on the IC to the aluminum heat sink. A small amount of IR radiation is emitted from the ICs and heat sink that you could see with an IR camera.

Heat sinks are rated in degrees centigrade/watt (°C/W). The specifications for the power ICs stipulate a heat sink rated at no more than 1.4°C/W, meaning that the temperature of the device attached to the heat sink shouldn't rise more than 1.4°C for each watt dissipated by the heat sink. This temperature differential is relative to the heat sink temperature, not the ambient temperature. For example, given 10W of dissipation and our 1.4°C/W heat sink rating, the ambient temperature may be 23°C (74°F), the heat sink temperature 35°C (95°F), and the device temperature of 35 + 14 = 49°C (120°F).

The AMP 100's heat sink is black anodized aluminum, weighs in at 19 ounces, with dimensions of about 7 3/4 × 2 × 2 1/2 inches. The °C/W rating for our heat sink is a function of the size and number of fins, type of aluminum, surface finish, surface area, and mass of metal. For example, the little corrugations on the fins increase the surface area and therefore the contact with cooling air, both of which decrease the °C/W rating. Even the color of the heat sink can affect the °C/W rating: a heat sink that is anodized flat black is a better thermal radiator than one finished in shiny aluminum.

All else equal, the larger the heat sink, the better the °C/W rating. In addition, the rating can be improved significantly by forced convection cooling, in which air is forced over the fins of the heat sink or chilled water or another coolant is circulated through the heat sink. A solid-state Peltier device, such as those used in battery-powered freezers, can be used to transport heat actively from the ICs to the heat sink or from heat sink to a larger heat sink.

But size and °C/W rating aren't the full story. If you bolt one of these power ICs to a 500-pound block of stainless steel, the average temperature of the block would increase very little, regardless of how hard you drive the TDA7294. However, the temperature of the stainless steel immediately surrounding the IC could rise considerably, even to the point of thermal shutdown.

The issue is thermal resistance, which roughly parallels the electrical resistance of materials. A high thermal resistance, like a high electrical resistance, impedes the flow of heat energy from the source. Silver has the highest thermal conductance of common metals, followed, in order, by copper, gold, and aluminum. If you've ever super-clocked a PC, you know that the better CPU coolers are made of copper for improved conduction over less-expensive aluminum coolers. As a point of reference, the thermal conductance of aluminum is ten times that of stainless steel.

If you removed the heat sink from the ICs during the teardown, you probably noticed that at the point of contact with the tab of the IC, the heat sink surface was smooth, bare aluminum. This maximizes the contact area and minimizes the thermal resistance between the tab of the IC and the heat sink. In addition, you probably encountered the messy thermal grease between the power ICs and the heat sink. The purpose of this thermal grease is to fill the microscopic irregularities in the two metal surfaces with a conductive media instead of thermally insulating air.

If you reconnect the heat sink to the ICs, use the smallest amount of conductive grease possible to cover the contact area. More is definitely not better, because the surplus grease diminishes the metal-to-mica contact area and thermal resistance skyrockets.

Because the tab on the TDA7294 is at –36VDC and the heat sink is bolted to the chassis, the IC must be electrically isolated and yet thermally bonded to the heat sink. This represents a challenge, because thermal conductors are generally electrical conductors. Mica is one exception. It's an excellent electrical insulator and a good thermal conductor. Hence the use of the thin mica sheets between the tabs of the IC and the heat sink, as shown in Figure 9-8. The thermal grease on both sides of the mica creates a thermally conducting, electrically insulating mica sandwich of sorts.

Alternatives to mica include polyimide, silicone rubber reinforced with fiberglass, and anodized aluminum. I've used these alternatives on 5W transistors, but I wouldn't dream of using silicone rubber, for example, on a device of more than 10W.

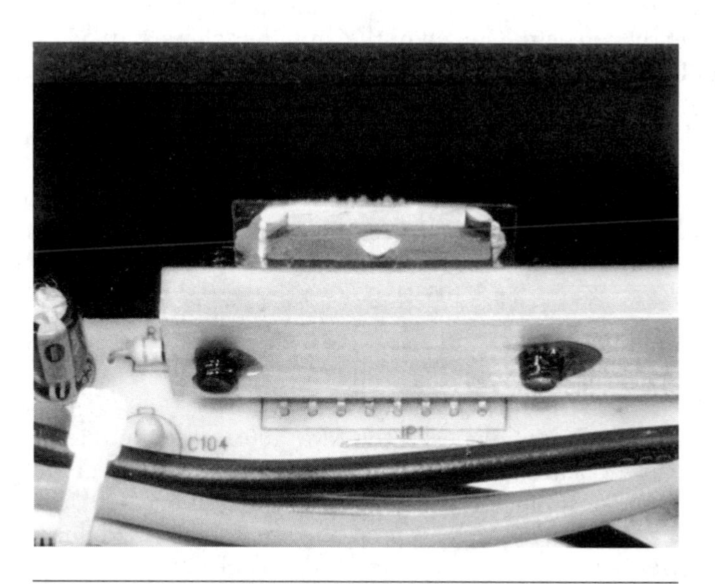

FIGURE 9-8 Mica insulator sheet between heat sink and amplifier IC

5532D and 4558D Operational Amplifiers

As mentioned during the teardown, the two other types of ICs used in the AMP 100 are 5532D and 4558D dual operational amplifiers. The 5532D is excellent for audio work because of its low noise and high bandwidth. The 4558D is a general-purpose operational amplifier, similar to the common 741, but with different pinouts. Both ICs are significantly less expensive than the power ICs. Digi-Key prices range from 15 to 30 cents for these chips in DIP packaging. Each 8-pin DIP chip contains two independent operational amplifiers.

C2235 NPN Transistor

The sole discrete transistor in the AMP 100 is a C2235, located on the logic board. This NPN transistor is rated at 120V maximum collector-emitter breakdown voltage, 800mA maximum collector current, and a current gain (Hfe) of 80. The TO-92 package transistor, capable of dissipating 900mW, is commonly used in audio power amplifier and driver stage amplifier applications.

Toroidal Transformer

The AMP 100's compact form factor is attributable in part to the massive, 62-ounce, low-profile toroidal power transformer that's a mere 1 1/2 inches thick and 4 inches in diameter, shown in Figure 9-1. A conventional, upright laminate power transformer with the same ratings could easily require twice the vertical clearance and nearly double the height of the amplifier chassis. In addition to a compact form factor, properly constructed toroidal transformers offer greater line noise immunity and higher efficiency, compared to conventional EI laminate core power transformers. (See the teardown of the ultrasonic humidifier in Chapter 7 for more information on EI core transformers.)

The toroidal transformer secondary delivers 26.8VAC center tapped, or 53.6VAC across the full secondary, at 3.2A. I measured the inductance of the primary winding at 2.8H and a DC resistance of 3Ω. At 60Hz, the inductive reactance of the primary is $2\pi fL$ or $2 \times \pi \times 60 \times 2.8 = 1055\Omega$.

Diodes

The most prominent diodes in the AMP 100 are the four 6A4 silicon power diodes on the power board. These diodes, which are the diameter of AAA batteries, are rated at 6A continuous, 400A peak surge current and 400V PIV (peak inverse voltage). As noted in the teardown, they're also elevated one body diameter above the power board to maximize convection cooling.

There are also four 1N4148 high-speed switching diodes on the switching board and one on the power board. These glass diodes, shown in Figures 9-6 and 9-7, are rated at 450mA and 100V PIV.

The most interesting diodes in the amplifier are the pair of 1N4744 zener diodes on the power board, shown in Figure 9-4. These glass diodes, rated at 15V and 1W, are used as +15VDC and –15VDC shunt regulators in conjunction with the 2W fusible resistors shown in the figure.

Capacitors

Most of the capacitors in the AMP 100 are electrolytic. As shown in Figure 9-3, there are two large 6800µf at 50VDC electrolytic capacitors in the center of the power board. The remaining 21 electrolytic capacitors, with values from 47 to 220µf at 25VDC, are distributed on the power board and logic board. Because the AMP 100 uses a dual supply system, the negative lead of some electrolytic capacitors are at ground potential, while the positive leads of others are grounded.

Two 0.1µf tantalum capacitors are associated with the potentiometers and two with the power ICs. Like electrolytic capacitors, tantalum capacitors are polarized. The half dozen ceramic disc capacitors are primarily low-voltage bypass capacitors. More prominent are the four 0.01µf and two 0.1µf Mylar capacitors used as bypass capacitors in the power supply and with the balance and volume control circuits. Finally, there are two rectangular 100nF metalized polyester film capacitors on the output of the power ICs, immediately adjacent to the audio output cable on the power board, as shown in the lower-right corner of Figure 9-5.

Resistors

A pair of 2W 470Ω fusible resistors is used in the zener diode shunt regulator circuit, shown in Figure 9-5, and a 1W 470Ω fusible is used in each audio output circuit, shown in Figure 9-3. Fusible resistors are designed to fail gracefully—that is, without flame, due to a flame-retardant coating—when subject to current above their specified power level.

The most striking resistors on the two circuits are the 1/8W metal film resistors. Instead of the usual 5 or 10 percent tolerance resistors, the boards are filled with 1 percent tolerance resistors. While the price differential between a 1 percent and a 5 percent resistor is probably a cent or less in large volume purchases, most manufacturers use the cheapest resistors available. The use of 1 percent resistors in the AMP 100 speaks to the quality of the amplifier. The value of these resistors can be a little difficult to decipher at first because of the extra digit band and because the multiplier band is brown, as opposed to the more easily identified gold or silver. Review the resistor color codes in Appendix A if you're not familiar with the five-band system.

Stacked Potentiometers

The four 100KΩ potentiometers in this amplifier are packaged as two stacked potentiometers with two potentiometers per shaft. Each shaft controls both left and right channels of the amplifier. The alternative, separate volume and balance potentiometers for each channel, would demand more front panel space and would likely put off most consumers.

Although they appear identical on the circuit board, there are mechanical and electrical differences between the two stacked potentiometers. A minor mechanical difference is the center notch on the balance potentiometer to indicate the midpoint of rotation. There is no mechanical notch on the volume potentiometer.

Electrically, the potentiometers are considerably different. The left and right balance potentiometers are linear taper, meaning the resistance between the center contact or wiper and one of the terminal contacts is proportional to the distance between the contacts and to the degree of rotation. You can expect the same resistance change per degree of rotation, regardless of the initial position of the wiper.

The volume potentiometer, in contrast, is a dual audio or log taper device, meaning the resistance change is a nonlinear function of the distance between the wiper and a terminal contact. The advantage of log over a linear taper is that the resulting change in volume more closely matches the logarithmic response of the human ear to changes in volume. Although the resistance change, and therefore output volume, is a nonlinear function of the potentiometer rotation, the change in volume is perceived as linear.

If you were to replace the log taper stacked potentiometer with the linear taper stacked potentiometer used for the balance control, the volume control would seem overly sensitive at low volumes and virtually nonresponsive at high volume settings. In fact, you should swap potentiometer positions and experience the effect firsthand.

Relay

The single relay in this amplifier, located on the right edge of the logic board in Figure 9-7, is a Goodsky GS-SH-212T. The DPDT relay has silver alloy contacts rated at 120VAC at 1A and 24VDC at 2A. The contacts close at about 17mA. Because the AMP 100 relies on silent convection cooling, the clicking is noticeable when the relay comes online.

As with manual switches, relay contacts are rated differently for AC and DC operation. In general, the AC specifications must be derated for DC operation. For example, if the Goodsky relay is used at 120VDC, the maximum current is likely around 250mA. (More detailed specifications on the relay are available at www.goodsky.co.uk.)

Fuse

The 4A slow-blow fuse is remarkable mainly in that it and the on–off push button switch are the only components between the power transformer primary and the power mains. There is no EMI filter or surge suppression, which I've come to expect in better electronics.

How It Works

A systems view of the AMP 100 is summarized in the block diagram in Figure 9-9. As shown in the figure, the amplifier consists of a logic board, a power board, a power transformer, a speaker selection switch, a power switch, and an output connector block. The source audio enters the system on the left, through the RCA input jacks on the logic board. The signal is buffered prior to being sent to a pair of drivers and then power amplifier ICs on the power board. While in the logic board, the signal is acted on by the sense circuitry, as a function of the stereo/mono and normal/auto switch settings on the logic board.

The power board contains the power supply, which is connected to the power switch module and the toroidal power transformer. It feeds the driver and power ICs on the power board and the sense circuitry and buffer on the logic board. The power board contains the volume and balance pots, accessible from the front of the AMP 100. The A/B speaker selection switch, accessed from the front of the amplifier, defines the electrical configuration of the output block in the rear of the unit. The output block is shown configured for stereo, with right and left speaker output. However, bridged output, using a single speaker between the left and right speaker output terminals, is also supported.

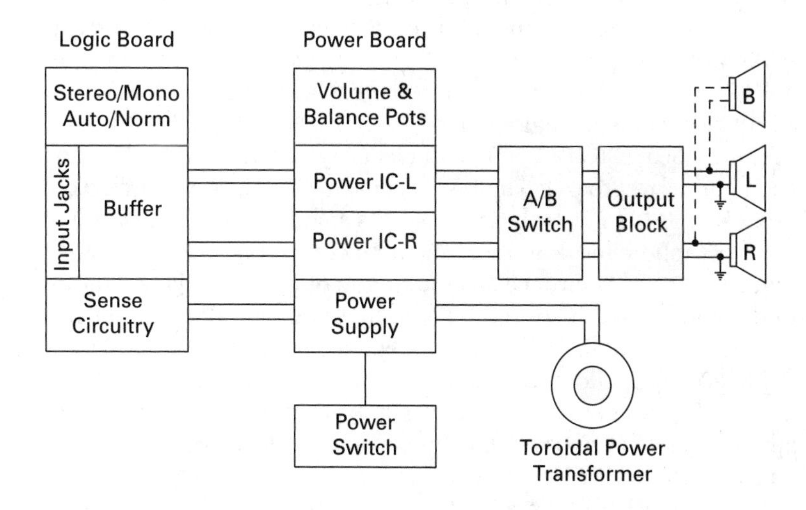

FIGURE 9-9 Block diagram of amplifier

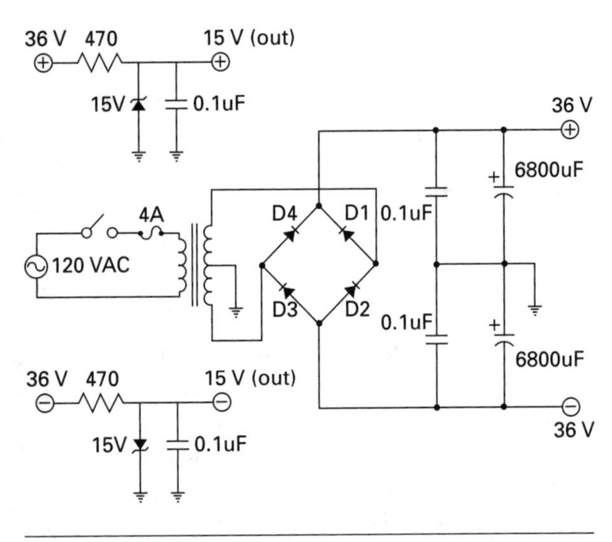

FIGURE 9-10 Schematic of power supply

Now let's examine the power supply schematic, shown in Figure 9-10. The 120VAC primary circuit is a bare-bones power switch and 4A slow-blow fuse. As I noted earlier, there are no inductors or filter capacitors in the primary circuit to minimize susceptibility to, or transmission of, EMI. The main power supply provides ±36VDC to the power ICs and the zener diode shunt regulators, shown at the top-left and bottom-left of the figure.

The secondary of the power transformer, which has a center tap to ground, feeds a bridge rectifier. In a standard bridge configuration, the negative junction of the bridge—in this case, the bottom corner of the bridge—is connected to ground, and the opposite, positive junction is connected to the electrolytic capacitors and other components of the output filter. However, in this power supply, both the negative and positive voltages output from the bridge rectifier are referenced to ground. As a result, instead of a single-ended 72VDC power supply, we have a dual-voltage supply that provides ±36VDC. A 6800µf electrolytic and 0.1µf Mylar capacitor are used on each output to ground. The Mylar capacitors shunt the high-frequency transients to ground, and the electrolytic capacitors smooth out the pulsating DC.

Let's follow the operation of the power supply through one cycle, using Figure 9-10 as a reference. When the top of the secondary winding is positive relative to the bottom of the winding, D1 and D3 conduct. Because the secondary center-tap is grounded, half of the secondary voltage to ground, or 36VDC, is applied to each of the 6800µf capacitors. On the next half cycle, when the top of the secondary winding is negative relative to the bottom of the winding, D2 and D4 conduct. Again 36VDC is applied across each of the 6800µf capacitors. Because this bridge conducts on every half cycle, it is a full-wave rectifier.

The ±36VDC is supplied directly to the TDA7294s. The operational amplifiers, relay, and transistor have much lower working voltages and are supplied by the two

15V zener shunt regulators. Unlike standard silicon power diodes that pop, crack, and fizzle when the peak inverse voltage is exceeded, zener diodes are designed to maintain a fixed voltage drop with reverse current flow. Within the current-carrying capacity of the two zener diodes, the voltage across each reverse-biased diode will be 15VDC, regardless of the applied voltage. In this regard, the zener diodes function as variable resistors. If the voltage across one of the 15V zener increases in magnitude, its resistance drops. If the voltage across the diode decreases, its resistance increases.

This increase and decrease in the resistance of a zener diode with changes in applied voltage can be used to create a voltage regulator when the zener is combined with a series resistor of suitable value—in this case, 470Ω. The role of each series resistor is to drop the 36VDC supply voltage to 15VDC and to control the flow of current through the zener diode. If the resistance of the series resistor is too low, the current capacity of the zener diode will be exceeded and the diode will fail. If the series resistance is too high, the zener diode won't conduct and the output voltage may swing wildly. The optimum value of the series resistor can be determined by Ohm's law:

$$R = V/I$$

where R is the resistance of the series resistor: in this example, 470Ω. V is the voltage across the resistor: in this example 36V – 15V = 21V. The sum of the zener current, I, is the current required by the zener diode to keep it in controlled reverse current breakdown mode, and the current required by the circuit. Given the 21V differential and 470Ω resistor value, total current requirement must be 21V/470Ω, or 45mA. The datasheet for the 1N4744A lists a zener current of 17mA. Assuming the circuit designers used the 17mA figure for zener current, the circuit load must be 45mA – 17mA = 28mA.

The advantages of a shunt regulator include simplicity and low cost. The disadvantages include poor regulation if the input supply voltage is poorly regulated and inefficiency. Using our figures for load and shunt current, the circuit load is 15V × 28mA = 420mW. The power dissipated by the 470Ω resistor is $I^2 \times R$, or $0.045^2 \times 470$ = 950mW. The overhead of delivering 420mW to the load is nearly 1W, which is dissipated as heat by each series resistor.

You might be wondering why use the zener diode at all. After all, if you know the circuit requirement is 28mA, why waste 17mA on a zener diode? If you had a constant load and a fixed supply voltage, you could get away with a series resistor. However, we can't make those assumptions here. Even if the steady-state current requirement of the circuit is 28mA, the startup current requirements are likely significantly greater. For example, during the first second or two of operation, electrolytic capacitors are uncharged and present a low impedance path to the initial voltage surge.

The second issue is variation in supply voltage, both at startup and during normal operation of the amplifier. The output of the nominal ±36VDC supply varies with changes in current drawn by the power ICs, as well as with fluctuations in the line voltage. If series resistors alone supply the ±15VDC, then the ±36VDC voltage fluctuations transfer directly to the ±15VDC output.

Now, for the analysis of the amplifier proper, refer to the simplified schematic in Figure 9-11. Let's follow the Line 1, right-side signal, starting with the RCA connectors on the logic board. A 5532D operational amplifier, configured as a voltage follower or buffer, leads to the relay controlled by the sensing circuitry. The buffer presents high input impedance and low output impedance, and provides unity gain. It isolates the signal source from the amplifier circuit so that each input channel to the AMP 100 essentially looks like the 47µf electrolytic capacitor and 100KΩ shunt resistor. This is an insignificant load for a typical signal source.

From a signal perspective, ideally the input would lead directly to the noninverting input of the 5532D operational amplifier. However, it's possible that the source could contain a DC offset voltage, which would skew the operation of the operational amplifier. The series capacitor ensures that the operational amplifier operates on an AC signal without a DC offset.

If you're hungry for more information on operational amplifier function, there's more detail in the teardown of an effects pedal in Chapter 12.

If there is a signal on Line 1, the relay defaults to the position shown in the schematic. The audio signal is sent through the miniature coaxial cable to the six-conductor cable assembly that connects to the right edge of the power board. Otherwise, if there is signal on Line 2 and no signal on the input of Line 1, the switching circuitry connects Line 2 to the power board.

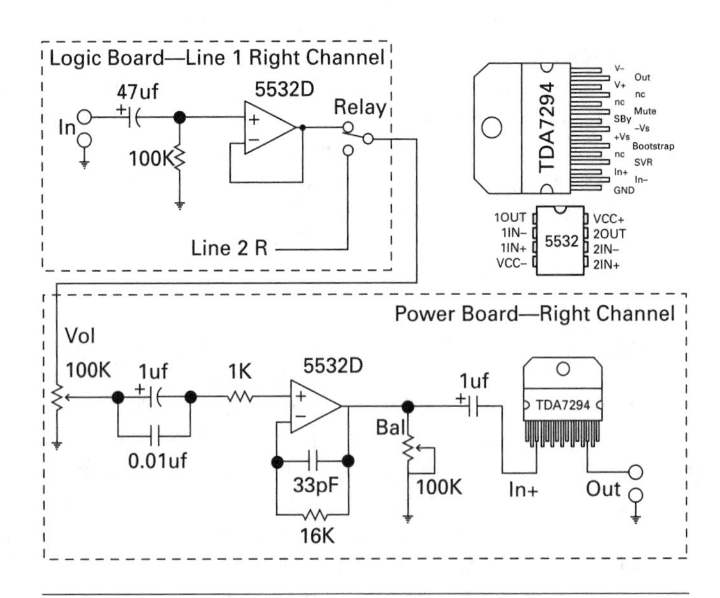

FIGURE 9-11 Simplified schematic of amplifier circuit

Note that, for clarity, the input circuitry for Line 2, which is identical to that of Line 1, is not shown in Figure 9-11. Similarly, in the following discussion of the right channel of the power board, unless stated otherwise, the circuitry and functionality are mirrored by the left channel.

Following the signal path from the logic board to the power board, the signal is sent to the volume potentiometer. The tap leads to the noninverting input of the only 5532D operational amplifier on the power board. This operational amplifier is configured as a noninverting driver, with the 45µf capacitor in the feedback loop to taper its frequency response.

The 5532D operational amplifier is followed by a balance control potentiometer, which acts a variable attenuator. Rotating the potentiometer shaft clockwise increases the right channel's resistance to ground while simultaneously decreasing the left channel's resistance to ground. The signal is then presented to the noninverting pin of the TDA7294 that handles the right channel.

Now let's analyze the logic board in more detail. This task is less arduous than it could be because the designers did a good job of separating the amplifier functions on the power board from the switching and logic functions on the logic board. Let's start with the stereo-mono switch and function, using the simplified schematic in Figure 9-12 as a guide. For clarity, the logic board circuitry is represented by a single operational amplifier and the stereo-mono switch, *SW*. Similarly, the left channel signal chain on the power board is represented by a volume control, driver operational amplifier, and power IC.

Recall that the function of the stereo-mono switch is to enable the AMP 100 to amplify the signal input to the right channel and either drive both speakers in

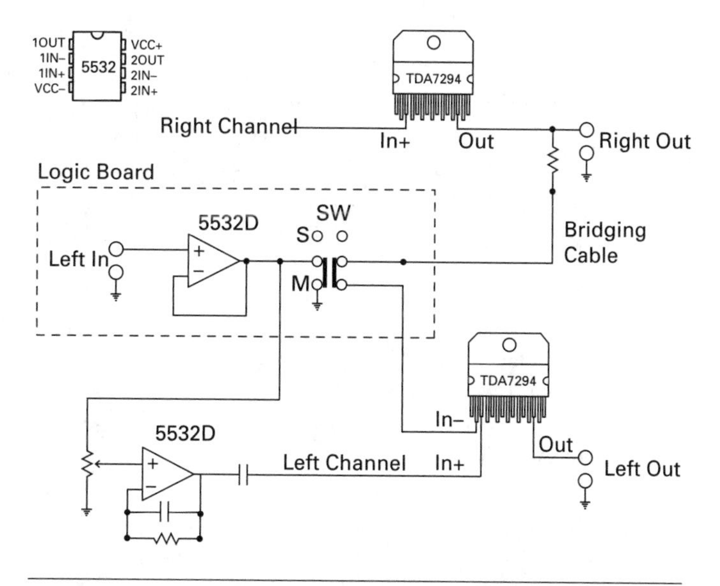

FIGURE 9-12 Simplified schematic of stereo-mono switching circuit

stereo mode or a single speaker in bridged mode. With the stereo-mono switch in the Stereo position, the switch is essentially out of the circuit. The signal input to the right RCA jack is sent to the noninverting (+) input to the TDA7294 dedicated to the right channel, and the signal input to the left RCA jack is sent to the noninverting (+) input of the other power IC. Both signals are subject to the same phase distortions. As such, if two signals are in phase at the input, they are in phase at the output of the amplifier.

When the stereo-mono switch is in the Mono position (M), as shown in Figure 9-12, the signal path is altered significantly. Now the output of the TDA7294 assigned to the right channel is sent to the inverting (–) input to the TDA7294 dedicated to the left channel via the coaxial bridging cable and series resistor. Any signal input to the left channel RCA jack is shunted to ground by the stereo-mono switch. This shunt essentially removes the remainder of the normal left channel signal chain from the circuit. Other than noise generated by the 5532D, no signal is presented to the noninverting (+) input of the left channel TDA7294.

The most important element of Figure 9-12 is the use of the inverting (–) input to the left channel TDA7294. This causes a 180-degree phase shift in the output of the left channel, relative to the output of the right channel. Recall that this phase difference is required for single speaker bridge amplification.

Now let's turn to the simplified schematic shown in Figure 9-13 to examine the circuitry associated with the normal-auto switch and the auto-input select function. As with the preceding schematic, circuit details, such as the components for biasing and operational amplifier feedback, have been omitted for the sake of clarity. The point is to understand the logic applied to signal flow.

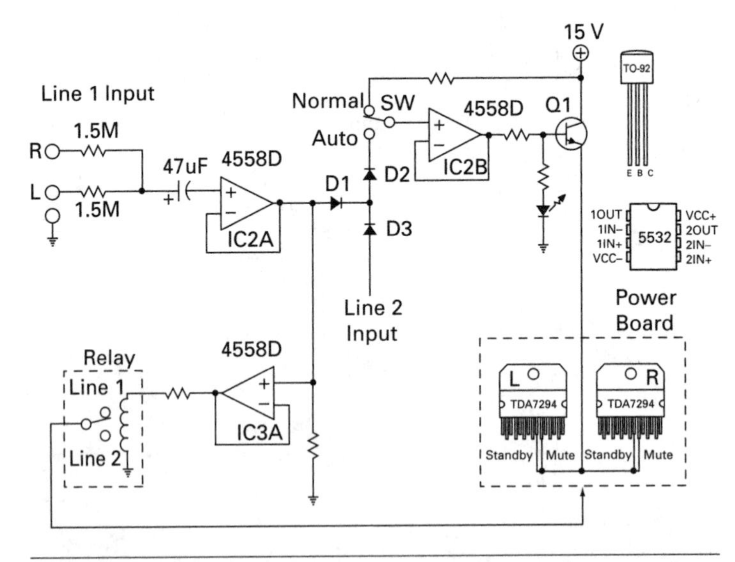

FIGURE 9-13 Simplified schematic of normal-auto and auto-input select circuits

Let's start with the function of the normal-auto switch. Recall that the position of this switch determines whether the front panel on–off switch toggles the AMP 100 between on and off or between on and active standby. With the switch (SW) in the Normal position, the noninverting (+) input of IC2B, a 4558D general-purpose operational amplifier, is tied to the positive 15VDC supply. This drives the output of the operational amplifier high, forward-biasing the C2235 NPN transistor and the LED connected to the base of the transistor and embedded in the front panel power switch.

The forward-biased LED simply indicates that the AMP 100 is on. The forward-biased C2235 transistor has a more significant role. As shown in the schematic, the emitter of the transistor is connected to the mute (*M*) and standby (*SBY*) pins of the two TDA7294 amplifier ICs. When the transistor conducts, the voltage on the pins is raised to about 12VDC, which deactivates the separate mute and standby functions, allowing the pair of TDA7294s to function as amplifiers. When the transistor is in the nonconducting state, these pins are at or near ground potential, which activates the mute and standby functions.

Now let's look at the circuit with the normal-auto switch in the Auto position. In this position, the amplifier is active when a signal is present on one of the input lines. Otherwise, the power ICs are muted and in low-power standby mode. From the preceding discussion, you should know that when the noninverting (+) input to IC2B is high, the LED is illuminated and the TDA7294s function normally. Conversely, when the input is low, the LED is dark and the TDA7294s are silent.

When a signal is present on either the right or left channel of Line 1, a signal is present on the noninverting (+) input to IC2A, a 4558D operational amplifier. The result is a positive output, which propagates through steering diodes D1 and D2 to the noninverting (+) input of IC2B. Similarly, when a signal is present on either channel of Line 2, the positive output propagates through steering diodes D3 and D2, to the noninverting (+) input of IC2B. In both cases, the LED and transistor are forward-biased.

Note that two 1.5M resistors are used to mix the inputs from the right and left channels of Line 1. Although the channel-to-channel resistance of 3.0M is significant, it represents a finite degradation in the separation of the two channels. An equivalent circuit is used for Line 2, which is omitted in the schematic for clarity. Of particular note is the use of switching diodes, D1–D3. Both Line 1 and Line 2 can contribute, independently, to the signal presented to input of IC2B. In addition, this isolation enables the auto-input select circuit to operate properly.

When a signal is present on Line 1, the output of IC2A is high. This signal drives the noninverting (+) input of IC3A, the output of which energizes the relay coil. The relay contacts are configured so that the right and left channels of Line 1 are fed to the power board.

Note that, if not for diode D1, the relay would be activated by a signal on Line 2. However, with D1 in the circuit, when a signal is present on Line 1 and not Line 2, the relay is not energized. The relay contacts route Line 2 signals to the input of the power board.

If you're following along at home with an AMP 100, you'll notice some components not included in Figure 9-13, such as a diode to protect IC3A from the inductive kick from the relay coil. If you're intent on modding the AMP 100, spend a few minutes mapping out the handful of additional components associated with the relay driver circuit. Also note the locations on the logic board where Line 1 and Line 2 are sampled for the switching logic.

Mods

It's hard to improve on a good thing, and this little amplifier is pretty good. However, you can address the most common complaints—erratic clicking of the relay as the auto-sensing circuit drops out during brief periods of low signal amplitude and lack of low frequency response—with the following mods.

Disconnect the Switching Circuitry

In theory, you could increase the SNR figure by disconnecting the sensing circuitry on the logic board. Simply unsolder one lead of each resistor that connects the sensing circuitry to the input connectors, using Figure 9-13 as a guide. Next, remove the resistor in series with the relay coil to avoid the annoying clicks sometimes present at startup. These steps will defeat the auto-input select feature as well as auto-on features, so you'll have to make due with manual on-off an a single input source. But the increased performance may be worth it to you. Fortunately, defeating the switching circuitry is an easy, readily reversible mod.

Create a Beefier Power Supply

If you follow the pro audio community online, you'll probably run across several references to lower than spec power output for the AMP 100 at 20Hz. To my ears, the existing circuitry is fine and I doubt that my speakers, headphones, or ears can handle anything below about 30Hz. Even so, if you want to feel your subwoofer rumble, the generic fix is beef up the power supply with additional filter capacitance.

There are three limits to this upgrade: the cost of high value, high density capacitors; the limits on additional capacitance imposed by the voltage rating of the power amplifier chips; and space. Give the limited height of the AMP 100's enclosure, you can't easily increase capacitance by installing taller capacitors of the same diameter. You might be able to locate higher density, higher capacitance replacement capacitors that fit on the board and in the enclosure, but they'll cost you. A more economical approach is to mount additional capacitors horizontally, adjacent to the two 6800µf capacitors.

If you really want to squeeze every bit of low-end response from the power ICs, consider moving the power transformer and power supply components to a separate enclosure. You'll have to feed the amplifier with ±36VDC and ±15VDC instead of 120VAC. If you opt for an external power supply, avoid the temptation of significantly increasing the capacitance of the filter circuit without modifying the circuit. If you double or quadruple the capacitance, you'll likely kick up the output voltage enough to require new series resistors for the zener shunt regulators. In addition to more capacitance, consider adding a high-current choke, and, if you have luck with locating surplus magnetics, separate transformers for positive and negative 36VDC supplies.

PART II

For Tinkerers

Chapter 10

Analog Volt-Ohm-Meter

The analog Volt-Ohm-Meter (VOM), such as the RadioShack VOM in Figure 10-1, has retained a place in the modern test instrument arsenal, despite the onslaught of inexpensive, mega-function digital multimeters (DMMs). For example, I keep both a Fluke 45 DMM and a Triplet 310 VOM on my workbench. The VOM, while not as precise or as accurate as the DMM, is my first choice for working with fluctuating or noisy signals and for rapid troubleshooting.

Not only is the response of my VOM virtually instantaneous, but the inertia of the meter movement serves as a mechanical low-pass filter to smooth fluctuations in the input signal. My five-digit Fluke 45, in comparison, often can't lock-in on a noisy signal. Even when my DMM is presented with a clean signal, it may take 5 or 6 seconds for the analog-to-digital converter and numerical display to settle down. So when accuracy and precision are called for, I turn to my expensive tabletop DMM. However, when I need speed and responsiveness, I turn to my analog VOM.

In this chapter, we'll tear down a RadioShack VOM, model 22-109, shown in Figure 10-1. The construction and feature set is typical of the sub-$20 instruments from companies such as Elenco, Mastech, and Mibuka.

Highlights

Unlike the typical, $10 single-chip DMM, this VOM offers a wealth of components and circuitry to examine. There is a mix of leaded and surface-mount components, including MOVs, Schottky, germanium, and general-purpose silicon diodes; a D'Arsonval galvanometer; a circuit board rotary switch; an electromagnetic buzzer; a linear pot; a fast-acting fuse; and, of course, resistors. This is a great teardown for brushing up on Ohm's law.

During the teardown, note the following:

- The circuit layout of the SMT components
- The marking system for SMT resistor value
- How to determine the tolerance of SMT resistors
- How you would improve the circuit design

FIGURE 10-1 RadioShack VOM

Specifications

The specifications for the RadioShack 22-109 VOM, hereafter referred to as simply *the VOM*, are typical for inexpensive but useful analog meters. Key specifications include the following:

- 0–500VAC/DC
- DC current from 0 to 250mA
- Resistance from 0 to 500K
- Audible continuity test
- Battery test feature

- Meter face is calibrated to display voltage readings in decibels (dB)
- Accuracy is ±5 percent of full scale
- DC sensitivity of 5000Ω/V

Significance

This is a good time to review the differences between accuracy, precision, and resolution as they relate to test equipment.

Accuracy is the closeness of the measured value to the actual or true value. An instrument is accurate if it indicates the true value. Accuracy can be affected by changes in temperature, drift in internal component values over time, and damage to the internal components from electrical or physical accidents and abuse.

Precision is a measure of dispersion or distribution around a given value, regardless of accuracy. It might help to think of precision as a measure of repeatability. If an instrument reads the gold standard source as 398.1VDC every time, then the meter has good precision—even if it's off the actual value by a couple volts.

Resolution is the granularity of measurement. It's more relevant to digital displays than to analog meters, because your perceptual abilities will limit the resolution with which you can read the position of the needle on the meter face. In the digital world, a measurement of 6.0001VDC has a greater resolution than a measurement of 6.0VDC. Resolution says nothing about accuracy or precision. A cheap DMM might indicate that a 400.00VDC source is 374.04857VDC—nice resolution, but the accuracy falls outside of the 5 percent full-scale accuracy figure of our inexpensive VOM.

The bottom line is don't be fooled by a string of digits after the decimal point. More numbers don't equate to more accuracy or more precision. For this reason, an expensive benchtop DMM that hasn't been calibrated in a while may be less accurate than an inexpensive, but recently calibrated pocket VOM. Unfortunately, however, our inexpensive VOM isn't equipped with a means of calibrating it against a calibration source.

Interpretation

According to my measurements, full-scale deflection of the VOM's galvanometer occurs at 200μA and 145mV. The DC sensitivity of the galvanometer, the reciprocal of the full-scale DC current, is 1/200μA, or 5000Ω/V. This figure agrees with the published specifications.

The greater the sensitivity and the truer the accuracy of a VOM, the better. To illustrate, my pro-grade (read expensive) Triplet 310 pocket VOM is rated at ±3 percent of full-scale accuracy with a DC sensitivity of 20,000Ω/V. The Triplet VOM is clearly a superior instrument—albeit at a significantly greater cost than the RadioShack VOM. However, these figures are unimpressive when compared with the specifications of my Fluke 45 benchtop DMM, which has a resolution of 1μV and accuracy of 0.025 percent. A caveat concerning DMMs is that the least significant digit is subject to rounding and digitization errors.

As with many inexpensive VOMs, our VOM has no AC amperage function. Overall accuracy is ±5 percent of full-scale, meaning a value displayed on the 500VAC scale should fall within ±(0.05 × 500), or ±25, volts of the actual value.

To put the specifications of these two VOMs and the DMM into perspective, let's assume a gold standard 400.00VDC source. Using the 500VDC scale on the RadioShack VOM, the reading should be 400 ± 25VDC. The reading could be as low as 375VDC or as high as 425VDC, and the VOM would still be operating within specification. The Triplet VOM, which has a 0–600VDC scale, should read 400 ± 18VDC because 3 percent of 600VDC is 18VDC. A reading of 375VDC would not be expected from the Triplet. The resolution of the two VOM meter faces is about the same. The Fluke DMM, in comparison, should read 400.00 ± 0.1VDC. The least significant digit might be off by a count or two because of limitations in the A–D converter, but that's a deviation of only a few hundredths of a volt.

Operation

If you're new to analog VOMs, the first thing you'll note is that it has no automatic features. When you plug in the test leads, make certain you plug the positive lead into the red, or positive, jack and the black lead into the black, or negative, jack. Unlike a DMM that will simply display a negative value when you test a negative voltage or current, with an analog VOM, you risk damaging the galvanometer and blowing the relatively expensive fuse when you don't observe proper polarity. In addition, some VOMs have jacks dedicated to specific functions, such as high current. These functions won't work unless the appropriate jacks are used.

Another key manual operation is selecting the operating mode—Resistance, DCV, ACV, Battery Check, DCmA, or Off—and the range within each mode. To maximize accuracy and repeatability, you should select a range that places the needle in the right third of the meter. For example, if you're troubleshooting a 120VAC circuit, you should use the 0–250VAC range instead of the 0–500VAC range. Recall that at an accuracy of ±5 percent of full-scale, a measurement using the 0–500VAC range is ±25VAC, versus ±12.5VAC for the 0–250VAC range. See Figure 10-2 for a close-up of the meter face with scale values. Note the nonlinearity and direction of the resistance scale, with higher resistance values compressed into a few graduations on the left edge of the scale.

You can maximize accuracy and repeatability of all measurements by using the parallax-correcting mirror embedded in the face of the meter. Position yourself so that the needle hides its reflection. If you do this consistently, you'll reduce parallax errors. Because of the space between the needle and the meter face, if you view the meter at an angle, the needle will appear to the left or right of the actual position over the meter face.

You can maximize the accuracy of resistance measurements by zeroing the needle prior to measuring the resistance of a component. With the meter set to resistance and the probes touching, adjust the zero pot so that the meter reads full-scale, or 0Ω.

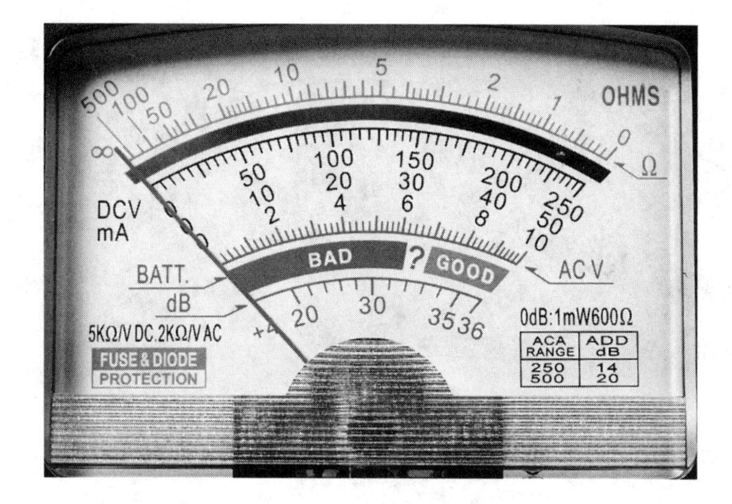

FIGURE 10-2 Meter face

You can make small changes in the resting position of the needle by adjusting the screw at the front of the meter with the VOM in the off position. This screw mechanically adjusts the galvanometer's spring position, and so the zero position of the needle.

Proper operation can also maximize accuracy. For example, a limitation of this and other galvanometer-based VOMs is susceptibility to magnetic fields generated by unshielded speakers and similar electronic equipment. If you need to take measurements in or around high-powered, unshielded systems, consider using either a shielded VOM or a solid-state DMM.

Teardown

As illustrated in Figure 10-3, this 5-minute teardown involves removing only a handful of screws. You'll have to handle the galvanometer with care once you extract it from the protective VOM enclosure.

Tools and Instruments

You'll need a Phillips-head screwdriver and a multimeter to trace the circuitry. If you decide to remove components for more detailed inspection, you'll need a soldering iron with a thin, clean tip; solder braid; and forceps.

Step by Step

If you want to test the functionality of the meter and practice reading the scale, do this before the teardown. If you're unfamiliar with reading resistance with an

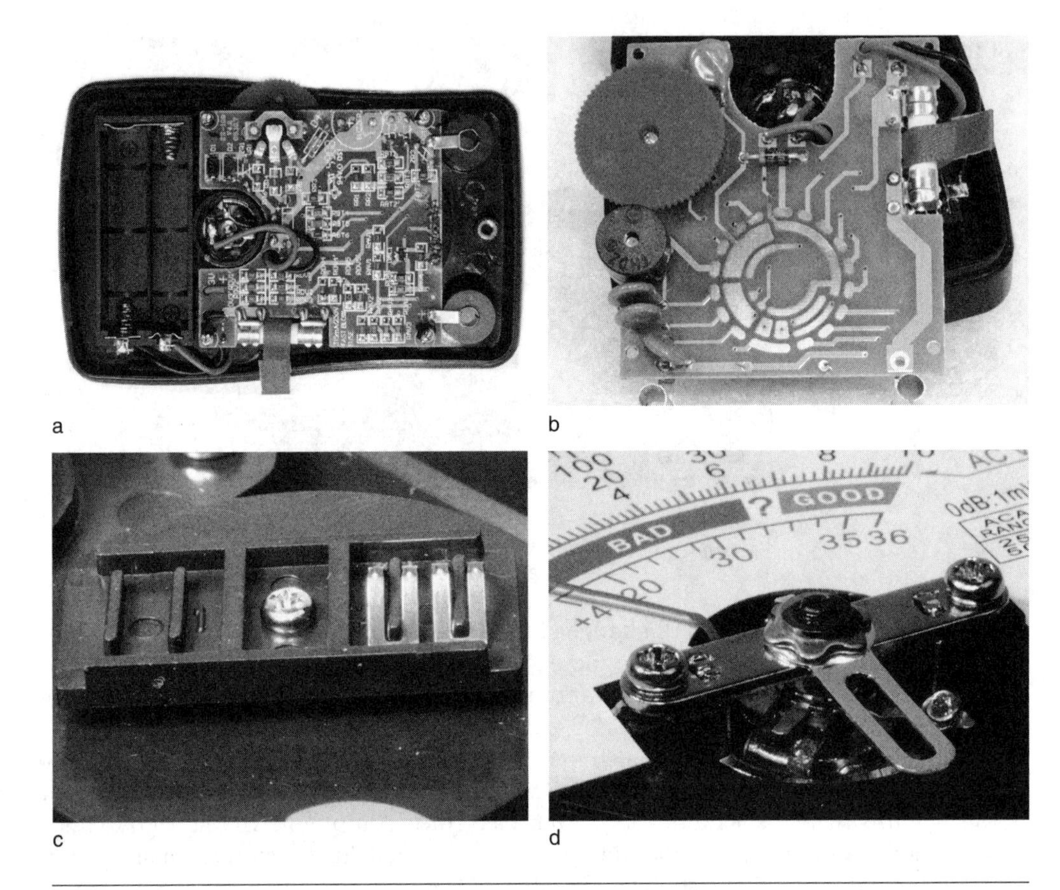

FIGURE 10-3 Teardown sequence

analog ohmmeter, for example, practice with a few resistors of known value. The more you understand the operation of the VOM, the more you'll appreciate the circuitry and components.

Step 1

Remove the back. Place the VOM facedown on a soft cloth or carpet remnant to avoid scratching the clear meter cover.

Step 2

Remove the two AAA cells, as shown in Figure 10-3a. Note the fast-blow fuse, the preponderance of surface-mount components, and the friction-fit positive and negative jack contacts.

Step 3

Release the printed circuit board, shown in Figure 10-3b. Note how the board forms part of the rotary switch. The matching rotary shorting bar is shown in Figure 10-3c.

Step 4

Extract the galvanometer, shown in Figure 10-3d. Pull the battery case to one side and then out, taking care to avoid damaging the needle.

Layout

The components of the VOM are laid out in quadrants. With the back of the VOM open and the empty battery holder oriented away from you, the components associated with DC measurements are primarily in the upper-left area of the board, adjacent to the fast-blow fuse (see Figure 10-4). The resistance components are in the upper-right, adjacent to the ohms zero pot. In the lower-left quadrant are the components associated with the AC measurements, and in the lower-right is the battery-check circuitry.

If you're following along at home with the same model of VOM, it's easy to identify areas devoted to specific meter functions because the components are labeled with circuit keys. For example, the prefix *RAV* is used with resistors in the

FIGURE 10-4 Component layout, top side, showing SMT resistors, fuse, and galvanometer

FIGURE 10-5 Component layout, bottom of circuit board

AC voltage section, *RR* for a resistor in the resistance section, *RDV* for a resistor in the DC voltage section, and so on.

As shown in Figure 10-5, the 17-position switch is integral to the bottom of the circuit board. Whereas the top of the board is populated predominantly with surface mount technology (SMT) components, the bottom of the board contains a handful of leaded components, including the large mini-buzzer, and a few leaded MOVs and a diode.

Components

As always, make a habit of looking up component data on the Web. Manufacturer-supplied datasheets often feature example circuit diagrams and other information that may not be evident in the teardown.

Precision Resistors

SMT resistors are the most numerous components in the VOM. The 1/8W resistors are packaged in 1206 cases (3.1mm × 1.5mm), and based on values, are 1 percent metal film. If you're looking for the familiar color bars to identify value and tolerance, you won't find them. Instead, numerical values are stamped on each package. An SMT resistor marked "7501" is "750" + "1" zero, or 7,500Ω. Similarly, a resistor marked "4993" is "499" + "3" zeros, or 499,000Ω.

Although tolerance values aren't marked, they can be easily deduced because resistance tolerance dictates the spacing of values. The lower the tolerance, the greater the spacing between successive values. For example, the progression of values in 5 percent tolerance SMT resistors is 39, 43, 51, and 56KΩ. The progression of 1 percent tolerance resistors includes 39.2 and 49.9KΩ values.

An import feature of SMT resistors is the maximum operating voltage. Common 1206 SMT resistors have a maximum operating voltage of 200VDC. Designers avoid arcing between SMT resistor solder pads and fried components in high-voltage circuits by using several SMT resistors in series.

The linear 15KΩ potentiometer, shown in Figure 10-6, is remarkable in that it's the only variable resistor in the VOM. Better VOMs have several pots that enable the manufacturer and users to calibrate the meter to compensate for variability in initial component values as well as drift in component values over time.

The third type of resistor used in this VOM, shown in Figure 10-7, is a 0.5Ω coil of solid nichrome wire. Nichrome, composed of about 60 percent nickel, with chromium, silicon, and iron for the balance, is commonly used for heater elements and air-wound resistors. Because the resistance per inch is known, creating a low-value resistor is simply a matter of using a specific length of wire where a resistor is needed.

21AWG nichrome wire has a resistance of 0.831Ω/ft. Higher gauge—that is, thinner—wire has greater resistance per foot but lower power-handling capability. The coiled shape does not affect electrical performance from DC to audio frequencies; I measured the inductance of the coil as only 2.4μH. However, a coil is a practical form to store a length of wire within the VOM.

FIGURE 10-6 15KΩ pot and SMT Schottky diodes

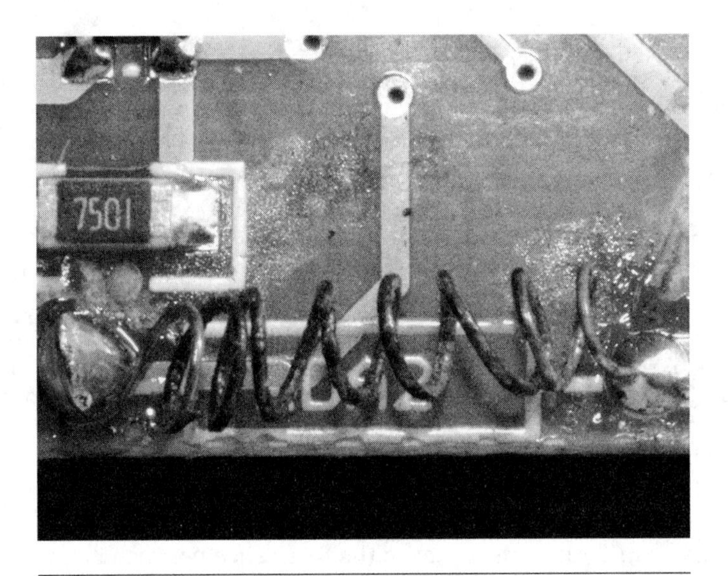

FIGURE 10-7 Nichrome wire resistor

Galvanometer

At the heart of every VOM is a galvanometer. These electromechanical displays rely on a magnetic field acting against spring tension to move a needle by an amount proportional to the current through a coil. The coil and spring of the galvanometer within the VOM are shown in Figure 10-8. This type of galvanometer is referred to as a *D'Arsonval movement*. The downward-pointing metal prongs, which engage with the screw in the meter face, are for adjusting the zero setting of the galvanometer.

As discussed earlier, the characteristics of the galvanometer determine the overall sensitivity of the VOM, and the more sensitive the better. According to my measurements, the resistance of this meter is 725Ω, full-scale voltage is 145mV, and full-scale current is 200μA. I made these measurements with my low-current bench DMM and a voltage divider circuit. If you try to measure the resistance of a galvanometer with an inexpensive VOM, you'll likely fry the galvanometer. The more expensive VOMs and virtually all DMMs use relatively low-current ohmmeter circuits.

Diodes

This VOM employs general-purpose silicon, Schottky, and germanium diodes. A general-purpose 1N4007 silicon diode protects the galvanometer when the VOM is in the off position, and a pair of Schottky SMT diodes protect the galvanometer

FIGURE 10-8 Galvanometer coil and spring viewed on axis

during measurements. The germanium diode protects the galvanometer against accidental polarity reversal in the battery check mode, and a pair of silicon diodes packaged as a single SMT component rectifies signals for AC voltage measurement.

The 1N4007, one of the few leaded components in the VOM, is rated at 1A and 700V PIV (peak inverse voltage)—the maximum reverse bias voltage that won't result in component failure. This silicon diode, visible at the far left of Figure 10-5, is switched across the galvanometer's positive terminal when the 17-position rotary switch is in the off position. Although the diode protects the galvanometer from damage if you accidentally touch the probes to a live circuit when the VOM is in the off position, it's a poor design guaranteed to sell fuses.

Better—and more expensive—VOMs not only disconnect the input from the leads in the off position, but they also short the leads of the galvanometer together. The short acts as an electromechanical brake that prevents the needle from swinging wildly if you drop the VOM or toss it into a toolkit. Place a clip lead across the galvanometer terminals and see how it dampens the movement of the needle.

The SMT Schottky diodes, shown in the top-right of Figure 10-6, differ from general-purpose silicon diodes in their forward voltage drop and speed of switching. A typical silicon diode, such as the 1N4007, begins conducting when the forward bias reaches about 0.7VDC. Schottky diodes, in comparison, begin conducting with a forward bias as low as 0.2VDC. Moreover, the switching time for a Schottky diode is orders of magnitude shorter than that of a typical silicon diode.

Considering the galvanometer in our VOM requires only 145mV for full-scale deflection, and that even a brief pulse of a significantly higher voltage could easily

FIGURE 10-9 Glass-cased germanium diode adjacent to potentiometer

fry the galvanometer coil, a typical silicon diode would be of little use. A limitation of Schottky diodes is their relatively low PIV ratings. However, because the diodes are wired back-to-back across the galvanometer, the diodes protect each other from reverse voltage spikes.

The glass-cased germanium diode, shown in Figure 10-9, is used in the battery check mode to guard against inevitable reverse-polarity mistakes. As with Schottky diodes, germanium diodes have a lower forward voltage drop than general-purpose silicon diodes—about 0.3VDC. This characteristic is useful, because inserting the germanium diode in series with the voltmeter circuitry will result in a battery check that's only 0.3VDC lower than the actual battery voltage. This is negligible compared to the 0.7VDC drop of a silicon power transistor. The use of a leaded germanium diode over an SMT Schottky diode probably reflects the higher cost of a Schottky diode.

Figure 10-10 shows the dual-diode SMT component (DAC1) used in the AC voltage measurement circuit. The three-terminal device contains two silicon diodes, arranged cathode-to-anode on one end and free cathode and anode at the other. It's not clear why this dual package was used over separate diodes, but component and assembly cost were probably factors.

Rotary Switch

The 17-position rotary switch is formed by a combination of the rotary shorting arm shown in Figure 10-3c and the conductive circuit board pattern shown in Figure 10-11. If you examine the backlit board in Figure 10-11, you can see how the

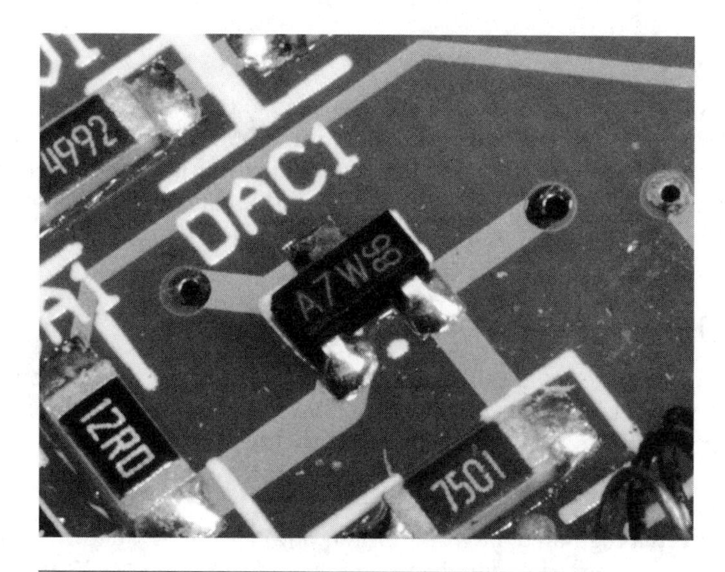

FIGURE 10-10 Dual-diode SMT package, DAC1

FIGURE 10-11 Rotary switch contacts, back illuminated

copper pads connect through conductive vias to traces on the component side of the board. This form of switch presents a modest obstacle to circuit tracing because the circuit board has to be mounted securely in place for the switch to function. Prepare yourself for removing and reattaching the circuit board a dozen times in the course of tracing the VOM's circuitry.

Battery

The two AAA cells merit discussion because their internal resistance is a component in the resistance mode of the VOM. Lithium, carbon-zinc, alkaline, nickel-metal hydride (NiMH), and nickel-cadmium (NiCd) batteries have different internal resistances that vary with the cell's state of charge. The ohmmeter features a zero adjustment pot to accommodate battery variability.

MOVs

Three unmarked, 0.3-inch diameter MOVs are used in the front ends of the battery check and ohmmeter circuitry to protect the galvanometer from electrical spikes. Recall from preceding chapters that varistors present a high resistance to low voltages and a low resistance once the applied voltage reaches a preset level.

I used a regulated, current-limited power supply to determine that the varistor voltage—the voltage at which current is 1mA—of these MOVs is 20V. At 24V, the current is about 10mA. Based on MOVs of similar diameter and varistor voltage in the Digi-Key catalog (www.digikey.com), the MOVs probably clamp input voltages to less than 30VDC and perhaps 20VAC. (For more information on evaluating MOVs, see Chapter 4 on surge protective devices.)

Fuse

The galvanometer is protected by a 315mA at 250V fast-blow fuse that's designed to fail before the galvanometer's fine copper wiring vaporizes. The speed of the fuse can be attributed to the use of extremely fine wire in the fuse, which unfortunately has a relatively high resistance of 3.9Ω. This resistance is significant at the lower resistance and current ranges of the instrument.

Electromagnetic Buzzer

The miniature electromagnetic buzzer provides audio feedback during continuity tests. The buzzer is a self-contained, high impedance (3.5MΩ) device with a built-in oscillator. According to my measurements, the device draws 20mA at 2V and 30mA at 3V. As shown in Figure 10-12, this buzzer has an outer magnet and an inner electromagnetic element.

The buzzer, visible in Figure 10-5, is the largest component attached to the circuit board. Even though piezoelectric buzzers tend to be more compact and lighter than electromagnetic buzzers, it's difficult to tell the two apart without a teardown. In place of the electromagnetic assembly, piezoelectric buzzers feature a thin disc of piezoelectric material that vibrates when excited by an oscillator built into the base of the buzzer.

FIGURE 10-12 Electromagnetic buzzer construction

Typical specifications for buzzers include operating voltage and current, operating frequency, and sound pressure level at a specified distance. The sound pressure specification for a typical electromagnetic mini-buzzer is around 75dB at 10cm. Recall that in this context, dB is calculated as follows:

$$dB = 20 \log_{10} (SPBuzzer/SPReference)$$

where SPBuzzer is the sound pressure created by the buzzer at a specified distance from the buzzer, and SPReference is the sound pressure reference, equal to the threshold for human hearing. Normally, dB is simply a ratio of two voltages, power, sound pressure, or other measurement. In this instance, dB is an absolute value.

A sound pressure level of 75dB is enough to cause you permanent hearing loss—not to mention headaches—with long-term exposure. Because the sound pressure level is inversely proportional to the distance from the sound source, you can save your hearing by not working directly over the buzzer. Simply moving the mini-buzzer 8 inches away from you would reduce the sound pressure level to a more bearable 69dB.

In case you didn't catch that sleight of hand, recall that doubling the power is equivalent to an increase of 3dB, and that doubling a voltage is equivalent to a 6dB increase. You can think of sound pressure as a voltage—both are pressures of sorts—and so the 6dB relationship applies. And don't forget that dB gain and loss are symmetrical, in that doubling or reducing the sound pressure by 50 percent involves a gain or loss of 6dB.

How It Works

The VOM is a combination of instruments that share a case, selector switch, and function-specific components. Let's examine each instrument function independently.

DC Voltmeter

The ideal DC voltmeter presents a high impedance to the circuit under test—the higher the better. Because a voltmeter is placed across a device or from a circuit element to ground, current flowing through the voltmeter can disrupt the circuit. The sensitivity of the galvanometer determines the current requirement for the voltmeter, which is another way of saying the instrument's input impedance. A 10M resistor draws considerably less current than a 10Ω resistor when placed across a signal source.

The galvanometer in our voltmeter requires 200μA for full deflection. In a 200W audio power amplifier circuit, a measurement that requires up to 200μA may be insignificant. However, in a 200mW preamp, a 200μA drain could shut down the circuit. Some analog VOM designs use a FET front end to minimize the current drawn from the circuit, but let's consider the simple case of a DC voltmeter without active elements.

Figure 10-13 shows the simplified DC voltmeter configuration of the VOM. The circuit consists of the 3.9Ω fuse, one or more series resistors, a pair of Schottky diodes, and the galvanometer. 725Ω is the equivalent resistance of the copper wire coil in the galvanometer. The multiplier resistors R1, R2, and R3, together with the resistances of the fuse and galvanometer, determine the voltage required for full-scale deflection.

As an exercise, let's determine the appropriate multiplier resistance for a 500VDC scale with our 200μA galvanometer. First, notice that the values of the switched multiplier resistors—R1, R2, and R3—are additive. The highest voltage range consists of R1, R2, and R3. The lowest range includes only R3. So keep in mind

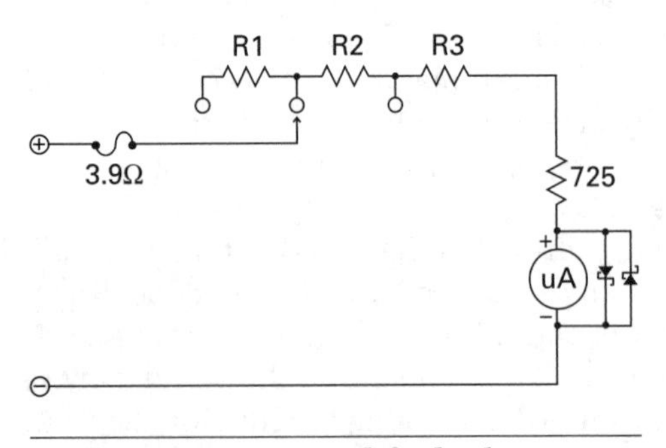

FIGURE 10-13 Simplified schematic of DC voltmeter

that when we solve for total resistance at the highest range, we constrain the possible values of R2 and R3. Let's assume that at the 500VDC range, the 3.9Ω contributed by the fuse and 750KΩ resistance of the galvanometer are insignificant.

Now, to Ohm's law:

$$R1 + R2 + R3 = \text{VFull-Scale}/\text{IFull-Scale}$$

$$R1 + R2 + R3 = 500\text{V}/200\mu\text{A}$$

$$R1 + R2 + R3 = 2.5\text{M}\Omega$$

Let's repeat the exercise with a full-scale voltage of 5VDC, and assign this to the middle switch position, which includes R2 and R4, as shown in Figure 10-13. Let's also use the 725Ω galvanometer resistance in our calculation.

$$R2 + R3 = \text{VFull-Scale}/\text{IFull-Scale} - 725$$

$$R2 + R3 = 5\text{V}/200\mu\text{A} - 725$$

$$R2 + R3 = 25{,}000 - 725$$

$$R2 + R3 = 24{,}275\Omega$$

Finally, let's assign the third switch position, which inserts R3 in series with the galvanometer. Let's use an arbitrarily low full-scale value of 0.5VDC to illustrate the significance of the resistance of the galvanometer, which is included in the calculation. Repeating the exercise:

$$R3 = \text{VFull-Scale}/\text{IFull-Scale} - 725$$

$$R3 = 0.5/200\mu\text{A} - 725$$

$$R3 = 2500 - 725$$

$$R3 = 1775\Omega$$

We can now determine the values of the three multiplier resistors, R1, R2, and R3, for full-scale values of 0.5, 5, and 500 VDC. We've determined the value of the 0.5VDC multiplier, R3:

$$R3 = 1775\Omega$$

We can calculate the value of the 5VDC scale multiplier, R2:

$$R2 + R3 = 24{,}275\Omega$$

$$R2 = 24{,}275 - R3$$

$$R2 = 24{,}275 - 1775$$

$$R2 = 22{,}500\Omega$$

Similarly, for the 500VDC scale multiplier, R1:

$$R1 + R2 + R3 = 2{,}500{,}000$$

$$R1 = 2{,}500{,}000 - R2 - R3$$

$$R1 = 2{,}500{,}000 - 22{,}500$$

$$R1 \approx 2{,}500{,}000\Omega$$

This exercise highlights the significance of resistor tolerances and the relative impact of fixed internal component values, such as the fuse and galvanometer, on the selection of multiplier resistance values.

One note in the overall design is that R1, R2, and R3 could be switched in circuit independently. The advantage of using the current design with several resistors in series with higher voltages is that the overall breakdown voltage is increased. Assuming a breakdown voltage of 200VDC per SMT resistor, three or more SMT resistors should be used in series on the 500VDC range of the VOM.

Battery Tester

The battery tester function of the VOM is essentially the DC voltmeter described previously with multiplier resistors selected for 9V and 1.5V full-scale, together with a dedicated, BAD, (unknown), and GOOD scale on the meter face (see Figure 10-2). In addition, the battery test circuit includes MOVs at each of the three switch positions and a germanium diode in series with the galvanometer. The MOVs absorb overvoltage spikes and the diode protects the galvanometer by insuring proper polarity of the battery under test. A powerful negative swing caused by testing a battery with the wrong lead polarity could permanently damage the galvanometer.

AC Voltmeter

If you compare Figures 10-13 and 10-14, you'll note several differences between the previously discussed DC voltmeter and the AC voltmeter. The most significant difference is a half-wave rectifier in the AC voltmeter used to convert an AC signal to a DC level. In addition, whereas the galvanometer in the DC voltmeter carries all current from the circuit under test, in the AC voltmeter, a proportion of the total current is shunted around the galvanometer.

When you measure a voltage with the AC voltmeter, current flows through the 3.9Ω resistor, through R3, and then to D1 and D2. When the top lead is negative, D1 shunts the current to the negative lead, bypassing the galvanometer. When the top lead is positive, current flows through D2 and the 7.5KΩ resistor in parallel with the galvanometer to the bottom lead.

Why not simply use one diode in series with the galvanometer? Because diodes are imperfect, and they allow a small leakage current when reverse-biased. Without

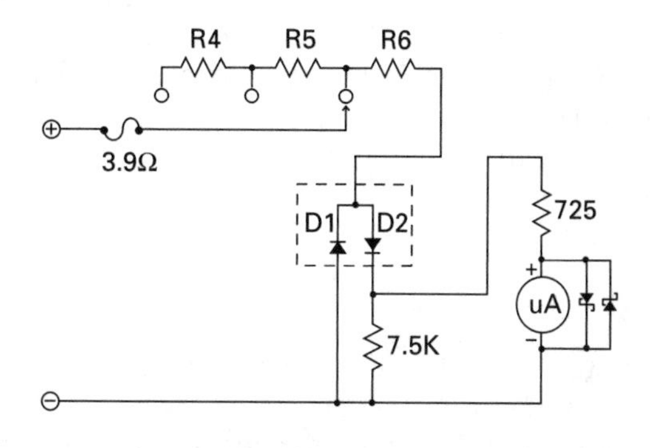

FIGURE 10-14 Simplified schematic of AC voltmeter

D1, some current would flow through the galvanometer when the top lead is negative. In a typical power supply circuit, a few hundred microamperes of leakage current are insignificant. In this circuit, that's enough current to vaporize the galvanometer.

Because the circuit uses half-wave rectification, the DC voltage at the junction of D2 and the 7.5KΩ resistor is a little less than half of the RMS (root mean square) AC applied across D1 and D2. For example, if we applied 120VAC line voltage across D1 and D2, we'd measure a little less than 60VDC across the 7.5KΩ resistor. The actual value of the voltage across the 7.5KΩ resistor, and therefore the galvanometer, isn't that important, as long as the meter face is calibrated to reflect the actual value. For example, if we ignore the 7.5KΩ resistor, we have a basic DC voltmeter with an error of about a half of a volt introduced by the forward bias voltage drop of D2. All we have to do is adjust the values of R4–R6 so that the desired full-scale voltage results in full-scale deflection of the galvanometer.

Are you curious about the purpose of the 7.5KΩ resistor? If you remove the resistor, you'll see that the voltmeter still functions, but the reading is a little off. Let's see by how much. We know full-scale deflection of the galvanometer requires 200μA, which is equivalent to 145mV across the 725Ω device. This voltage would result in a current flow through the 7.5KΩ resistor of the following:

$$I = V/R$$

$$I = 145\text{mV}/7.5\text{K}\Omega$$

$$I = 0.145\text{V}/7500\Omega$$

$$I = 19\mu\text{A}$$

So about 10 percent of the current through D2 passes through the 7.5KΩ resistor, and 90 percent flows through the galvanometer. Why would the designers intentionally decrease the effective sensitivity of the voltmeter? One reason would be to use standard 1 percent resistor values for R4–R6. Recall that we can't use

R1–R3 from the DC voltmeter circuit because of the voltage drop from D2 and the half-wave rectification equivalent of a little less than half (approximately 45 percent) of the RMS AC voltage. Better meters use variable resistors in series with the galvanometer circuit to apply the proper correction. Can you think of another reason for the 7.5KΩ resistor?

Ammeter

The ideal ammeter is a low-resistance device that is placed in series with the circuit under test. Our ammeter circuit, shown in Figure 10-15, illustrates many of the challenges associated with achieving a low-resistance instrument. First, let's ignore the 0.5, 12, and 20Ω shunt resistors and focus on the leads, fuse, and galvanometer. The maximum current we can measure with our bare galvanometer is 200µA. At this level of current, the resistance of the test leads and fuse are insignificant.

Now let's say we want a full-scale current of 250mA. We have to shunt all but 200µA of that current through a low-resistance device in parallel with the galvanometer, as shown in Figure 10-15. But how do we calculate the value of the shunt resistance? The easiest way is to recall that full-scale deflection occurs with 145mV across the galvanometer. So we have this:

RShunt = VFull-Scale/IFull-Scale

RShunt = 145mV/250mA

RShunt = 0.58Ω

I measured the resistance of the nichrome wire shunt used in the 250mA range as 0.5Ω, which is close to the calculated value.

Similarly, for 25mA full-scale, the shunt resistance should be as follows:

RShunt = 145mV/25mA

RShunt = 5.8Ω

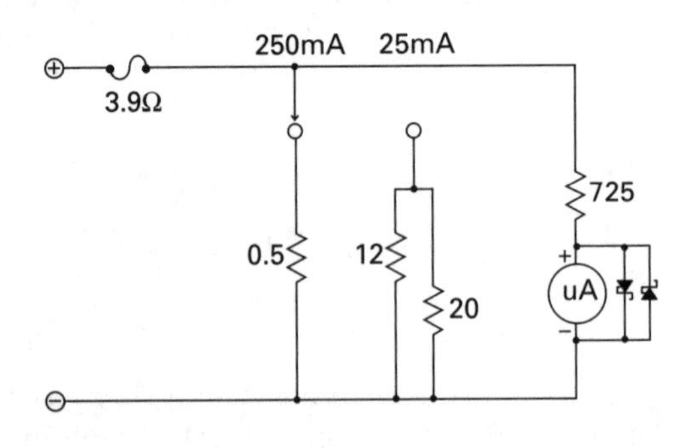

FIGURE 10-15 Simplified schematic of ammeter

Because 5.8Ω isn't a standard 1 percent SMT resistor value, the designers used a 12Ω and 20Ω resistor in parallel for an equivalent resistance of

$$RShunt = (12\Omega \times 20\Omega)/(12\Omega + 20\Omega)$$

$$RShunt = 7.5\Omega$$

Another reason to use multiple resistors in parallel in a current shunt is to increase the dissipation capabilities of the shunt. Two approximately equal 1/8W resistors can dissipate 1/4W when connected in parallel.

Notice that in both circuits, I've ignored the contribution of the 750Ω equivalent resistance of the galvanometer and assumed that most of the current flows through the shunt resistor. Even at 25mA full-scale, 200μA or 0.2mA is insignificant. The difference between the calculated and actual value of the 25mA shunt used in our VOM is significant, however. Again, the rationale for using approximate values is probably a reflection of component cost.

Another significant aspect of the circuit is the resistance of the fuse. On the 250mA scale, the ammeter inserts about 4.5Ω in series with the circuit under test. Assuming a 12V circuit, at 250mA, the voltage drop across the VOM is 4.5Ω × 0.25A = 1.1V. This represents about a 10 percent change in the operating voltage of the circuit, enough to misrepresent the actual current drawn at the full operating voltage. A transistor amplifier circuit might draw less current at a lower operating voltage, but a small DC motor might draw more current, depending on its electrical configuration and mechanical load.

Ohmmeter

The ohmmeter component of our VOM is a little more complicated than either the voltmeter or ammeter, as illustrated by the simplified schematic in Figure 10-16. However, if we ignore the protective elements of the circuit, we have a basic DC ammeter circuit with a built-in current source. The diode D1 protects the meter from reverse voltage resulting from an improperly installed battery—at the cost of a blown fuse. In addition, the MOV protects the meter from inevitable voltage spikes resulting from, for example, checking the in-circuit resistance when there are partially discharged capacitors.

To get a handle on circuit operation, let's make some simplifying assumptions. We can assume that the leads are shorted together, so that R = 0Ω. We can ignore the contributions of the 1MΩ resistor and 3.9Ω fuse. We can assume that the 15KΩ pot is centered and presents a 7.5KΩ resistance in series with the galvanometer. With these assumptions, we have a 3VDC source and a 47Ω resistor in parallel with the galvanometer circuit with a resistance of 4.9KΩ + 7.5KΩ + 4.3KΩ + 725Ω = 12,525Ω. In this configuration, current through the galvanometer is as follows:

$$I = V/R$$

$$I = 3V/12{,}525\Omega$$

$$I = 240mA$$

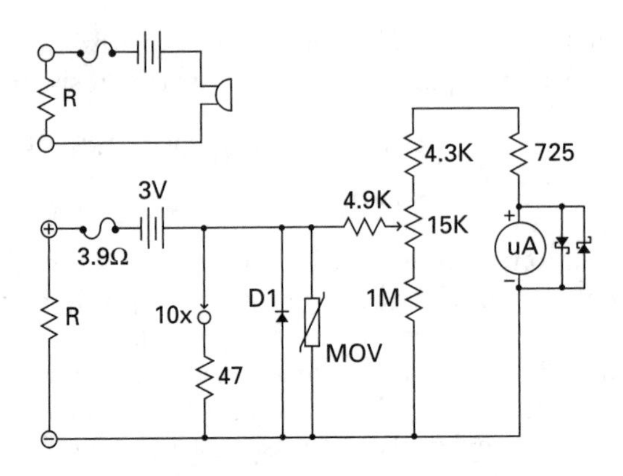

FIGURE 10-16 Simplified schematic of ohmmeter in X10 position (center) and continuity test circuit (top-left)

This is greater than the full-scale value of 200μA, but about right. We can increase the series resistance contributed by the 15KΩ pot so that the total resistance is 3V/200μA = 15KΩ. In actual use, you would adjust the pot to compensate for the contribution of the fuse resistance, the actual battery voltage, as well as the internal resistance of the battery. Recall that the internal resistance of the battery affects current flow according to Ohm's law, and that this resistance depends on the charge status of the battery and the battery chemistry. Lithium batteries have a lower internal series resistance than carbon-zinc batteries, for example.

Now let's see what happens when R = 100Ω. If you look at the meter face, you can see that, on the 10X setting, 100Ω corresponds to about 30 percent deflection of the galvanometer, or a current of about 65μA through the galvanometer circuit. Keeping this in mind, let's return to the schematic in Figure 10-16. Now we have a 3V source, a series resistance of 100Ω, and the 47Ω resistor in parallel with the 15KΩ equivalent resistance of the galvanometer circuit. What's the current through the galvanometer?

We can ignore the 15KΩ resistance and determine the voltage across the 47Ω resistor in series with the 100Ω resistor. Let's start by determining the total current and then calculating the voltage drop across the 47Ω resistor:

$$I = 3V/(100\Omega + 47\Omega)$$

$$I = 20mA$$

Now, let's calculate the voltage across the 47Ω resistor:

$$V = I \times R$$

$$V = 20mA \times 47\Omega$$

$$V = 0.94V$$

Assuming an equivalent series resistance of 15KΩ, current through the galvanometer is

$$I = V/R$$

$$I = 0.94V/15{,}000\Omega$$

$$I = 63\mu A$$

This is fairly close to our original guesstimate, which was based on the meter face markings.

The current through the galvanometer is a function of the ratio of the unknown resistance to the resistance of the 47Ω resistor, and not the absolute value of the unknown resistance. This ratio measurement explains why the scale is compressed at the higher resistance levels. In terms of galvanometric current, a change in the resistance ratio from 1:1 to 2:1 is significant. However, the difference between 10:1 and 11:1 is barely perceptible.

To illustrate, let's use the meter face to estimate galvanometer current. On the X10 setting, testing a 47Ω resistor should place the needle midpoint in the scale, or about 100μA. With a 100Ω resistor, the deflection is about a third, corresponding to 200μA/3, or about 66μA. At 500Ω, deflection is only about 10 percent of full-scale, or 20μA, and so on. Above 500Ω, it's difficult to discern deflection and corresponding galvanometer current.

Keep in mind that the value of the shunt resistor depends on the meter setting. At X100 and X1K, the shunt resistors are 523Ω and 7.5KΩ, respectively. Using a shunt resistor near the value of the unknown resistance results in a galvanometer deflection near mid-scale, away from the compressed, difficult-to-read left edge of the scale. You can use this feature to your advantage if you work with components with a specific range of resistances.

For example, let's say you use your VOM on an assembly line to test guitar pickups, which should have a resistance of 10–12K. If you replace the 47Ω resistor used in the X10 setting with a 20KΩ, 0.1 percent tolerance resistor, you could test coil resistance with the needle in the right third of the scale where small differences in resistance are discernable. You could mark the meter face for correct resistance values, or simply make a note to yourself that components should test within, for example, 1.5 and 2.0 on the existing resistance scale.

Audible Continuity Tester

The audible continuity test mode provides a fast, qualitative test for resistance. As illustrated in the upper-left of Figure 10-16, the unknown resistance simply completes the circuit comprising the 3.9Ω fuse, the battery, and the buzzer. My measurements indicate that the buzzer begins working when it draws about 20mA—which translates to a maximum of 3V/20mA = 150Ω between the test leads. This contrasts with the stated minimum resistance of 300Ω. The continuity tester is great for checking grounds and circuits that don't involve sensitive electronic components.

This VOM is fine for testing diodes, capacitors, resistors, and some fuses, but not for components such as a 200μA galvanometer. The continuity tester essentially places the 3V battery across the component. In our galvanometer, this would result in a brief current of 3V/750Ω = 4000μA. The X10 and X100 settings of the ohmmeter would result in essentially the same current flow. Even the lowest current setting, X1K, would result in a current of 3V/(7500Ω + 725Ω) = 365μA through the galvanometer—more than enough to fry its coil. The ohmmeter function, like the other functions on this VOM, is valuable and useful on and off the workbench. As with any test instrument, you simply have to appreciate the strengths and limitations of the tool.

Mods

You can improve upon the 5 percent full-scale accuracy of the VOM by replacing the 1 percent resistors with 0.1 percent resistors. At about 25 cents per SMT resistor, it should cost about $8 to replace all 32 SMT resistors. In return for your efforts, you might realize a 1 percent improvement in accuracy.

A more practical mod, or perhaps application, of the VOM is to create a high-current meter. If you want to measure currents up to 10A with your VOM, you can add an external shunt resistor and measure the voltage across the shunt using the 10VDC scale. So what's the value of the shunt resistor? Assuming we use the VOMs 10VDC scale, from Ohm's law: R = 10V/10A = 1Ω. The minimum power dissipation required is P = 10V × 10A = 100W.

So if we wanted to use the 10V scale, we'd either need a 1Ω, 100W resistor or some combination of lower wattage, less expensive resistors. In practice, a 150W resistor or resistor network is more practical because of the safety margin. Figure 10-17 shows an external current shunt composed of 20W resistors connected in parallel that I picked up for $5 on eBay.

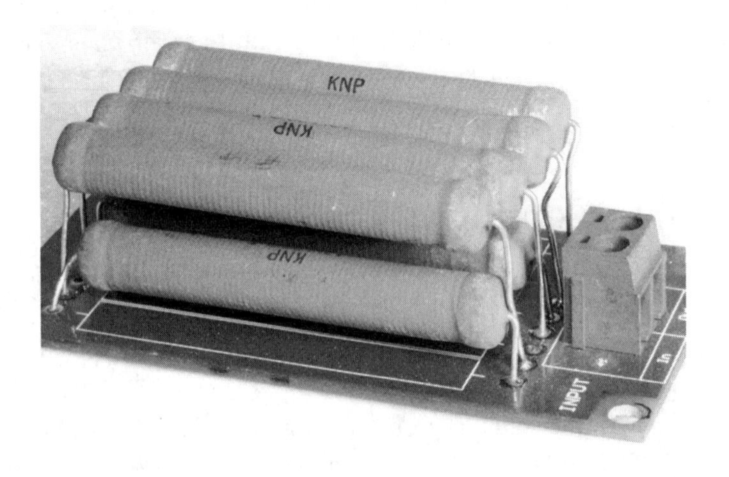

FIGURE 10-17 External current shunt

If you decide to build your own high-current shunt, you should buy a few more resistors than necessary. Because of resistor tolerance, your final shunt could be off by a tenth of an ohm or more. The solution is to use your ohmmeter to select the combination of resistors that will provide a shunt as close as possible to 1Ω.

The power requirements for the shunt illustrate the advantage of a more sensitive VOM. For example, a 20,000Ω/V Triplet 310 has a 3VDC scale. Even though you'd have to multiply the meter reading in volts by 3.33 to get the actual current in amps, you could get by with a 0.33Ω resistor: R = 3V/10A = 0.33Ω. Unfortunately, the power rating would be significant: P = 3V × 10A = 30W. However, with a lower value external shunt, the power requirement is lower as well. This translates to cost savings and, more importantly, to less of an impact on the circuit under test.

Chapter 11

Laser-guided Sonic Distance Measurer

I own three tape measures—a figure that's probably well below the national average. I have a compact, 6-foot tape measure near my workbench to measure chassis layouts, to get a quick estimate of circuit board sizes, and, in general, to work with small objects. Then there's the 100-foot monster tape, with a built-in crank handle, for measuring outdoor cable runs, spacing between trees, and other odd surveying tasks. Finally, I have a 25-foot tape that I use for measuring carpet, walls, and large interior spaces that the 6-foot tape just can't cover. The problem with the 25-foot and 100-foot tapes is that I generally need someone to hold the other end. Tapes droop unless they're supported by the floor or they simply dislodge because there's no one to ensure that the free end is secured.

Several technological alternatives to the inexpensive but sometimes limited tape measure are available, including handheld ultrasonic and laser distance measuring devices. Ultrasonic distance measurers, from companies such as Zircon (see Figure 11-1), Strait-Line, RadioShack, Mastech, and Black & Decker, are relatively affordable at $10 to $80. Sonic measurers, also called ultrasonic tape measures, are based on an ultrasonic transceiver that emits an ultrasonic burst and then times the return. Most of these sonar-like devices have a visible laser for a pointing aid. Sonic measurers for home use operate around 40kHz and generally offer accuracy within a few inches at the maximum range of 50 feet.

Laser distance measuring devices time the round-trip time of laser light instead of an ultrasonic signal. As a result, their maximum range is typically 250 feet when used with a special reflective target, and accuracy is significantly better than with an ultrasonic unit. Prices for nonprofessional laser units from companies including Bosch, Mastech, and Strait-Line range from about $50 to $200, however. Too expensive for my tool chest, especially when a good 100-foot tape measure sells for about $20.

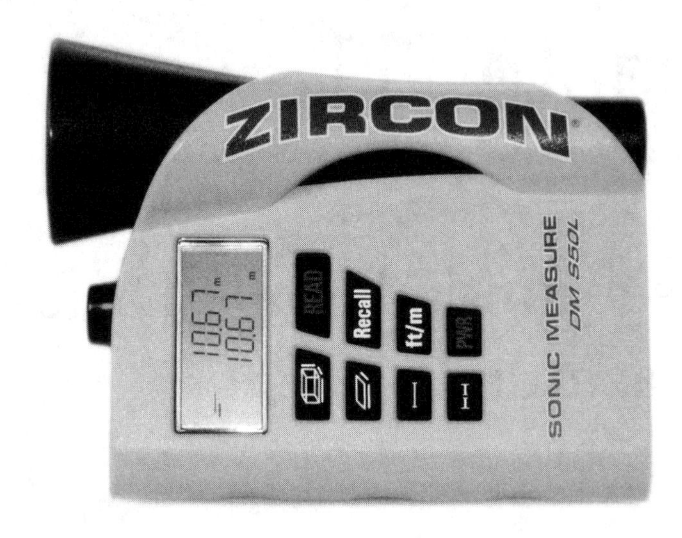

FIGURE 11-1 Zircon sonic measurer

In this chapter, I'll tear down the Chinese-manufactured Zircon Sonic Measure DM S50L, shown in Figure 11-1. You can buy this laser-guided ultrasonic distance measurer, hereafter also referred to as the "sonic measurer," from Amazon for about $40. This teardown provides an excellent means of exploring the physics of sound and an example of interfacing analog sensors and effectors with a microcontroller. At the end of the chapter, I'll suggest a few ways to repurpose some of the circuitry.

Highlights

This is a fun, relatively short teardown. Although the circuit board features a mysterious processor hidden under a blob of black epoxy, we have plenty of discrete components to examine, including an ultrasonic transducer, a thermistor, two voltage regulators, and a visible laser diode assembly.

During the teardown, note the following:

- The layout of the circuit board, which is primarily populated with SMT components
- The bulk of the laser diode module
- The ultrasonic transceiver driver circuitry
- The power regulation and distribution system

Specifications

The specifications of the Zircon DM S50L, extracted from the one-page user guide, the Zircon web site (www.zircon.com), and the back of the unit, include the following:

- **Range:** 2–50 feet
- **Resolution:** 1 inch
- **Accuracy:** ±0.5 percent of distance ±1 digit
- **Operating temperature:** 20°F–120°F
- **Operating humidity:** 0–80 percent (noncondensing)
- **Functions:** Area, volume, addition, memory recall
- **Power requirements**: 9V alkaline battery for 8 hours of continuous use; auto-off after 30 seconds
- **Aperture angle:** ±5 degrees from center
- **Laser:** Class IIIa, <5mW at 630–670nm; compliant with 21 CFR parts 1040 10 and 11
- **Dimensions:** 4 3/4 × 3 1/2 × 1 2/3 inches (HWD)
- **Weight:** 4.7 ounces

These specifications are useful in comparing units from different manufacturers and for insight into the inner workings of the sonic measurer. Let's consider the more significant specifications of the unit in more detail.

Range

The range of the sonic measurer is a function of the properties of ultrasound, the design of the ultrasound transducer, the receiver circuit, and the available processing power. The minimum distance reflects limitations in the processor and transceiver design, which in turn reflect practical cost constraints. The shorter the distance from transceiver to the wall or other object, the quicker the transceiver and processor must respond to the echo. Because the ultrasonic transducer operates in a simplex mode—it can either send or receive, but not both simultaneously—a refractory period follows transmission, during which the transducer cannot receive an echo.

However, there's nothing inherent in the design of a single-transducer transceiver that prevents a minimum range of less than 2 feet. For example, a single-element ultrasonic rangefinder that's popular in the robotics community, the Devantech SRF02 Sensor (www.acroname.com), has a minimum distance of 6 inches. The SRF02 achieves this minimum distance through fast recovery and onboard signal processing. However, the cost of the SRF02 is about $25.

Another approach to minimizing the working distance of an ultrasonic measurer is to use separate transmitter and receiver transducers, separated horizontally by a centimeter or two. For example, the Devantech SRF04 Ranger offers a minimum measurement distance of about 1 inch, in part because the receiver element can be active at all times. However, while a dual-transducer transceiver may be fine for a large robot, there's no space on a handheld unit for two horn assemblies—one for transmit and one for receive.

The maximum range of an ultrasonic measurer is a function of the effective radiated power of the transmitted ultrasound chirps, the sensitivity and selectivity of the receiver, the nature of the target and echo, and the operating frequency. The greater the effective radiated power, the greater the maximum range, everything else held constant. Effective radiated power can be increased by increasing power to the transducer, by using a more efficient transducer, and by employing a horn or acoustic antenna that produces a narrower, more focused acoustic beam. The more focused the transmitted signal, the greater the energy imparted to the target and reflected. A solid, flat, broad surface, perpendicular to the horn, produces the best echoes.

The greater the sensitivity of the receiver and the more focused or selective the horn assembly, the more likely an echo will be detected, assuming perfect target alignment. However, because the target may not be precisely perpendicular to the transmitted signal, a horn that's too narrowly focused may reject the echo. So the directivity of the horn must be low enough to allow the receipt of off-axis signals. The catch is that off-axis reflections from objects other than the intended target can also make it to the receiver if the directivity is too low. Because the transmitter and receiver share the same horn, the directivity of the horn is a compromise between narrowly focusing the transmitted ultrasound signal presenting an aperture wide enough to capture an off-axis echo.

The frequency of the ultrasound signal significantly constrains the maximum distance because of differences in the propagation and of sound through the air as a function of frequency. For example, ultrasonic rangefinders operating at 235kHz have a number of advantages over rangefinders operating at 40kHz, such as a potentially tighter transmitted acoustic beam. However, whereas a sonic measurer based on 40kHz sound can measure up to 50 feet or more, an equivalently configured 235kHz rangefinder, such as the Devantech SRF235 (www.acroname.com), is limited to about 47 inches, because the higher frequency signal is attenuated by the air.

Resolution

If you read the Volt-Ohm-Meter (VOM) teardown in Chapter 10, you know that *resolution* refers to the number of significant digits that can be displayed by the meter. Similarly, resolution in this case is simply the finest level of granularity supported by the sonic measurer—about 1 inch. Resolution is a function of the numeric capabilities and programming of the microcontroller and the number of digits supported by the LCD module.

Accuracy

Also covered in the VOM teardown, accuracy reflects the trueness of the measurement. It's a function of the algorithms used in the microcontroller, rounding errors due to hardware limitations, the accuracy of the microcontroller's timebase, provision for environmental variability, the accuracy of the factory calibration, the nature of the target, and user error. In addition, because components shift value with age, time since calibration is also a factor.

At 50 feet, the claimed accuracy of the sonic measurer is within ±0.5 percent of 50 feet ±1 digit, or ±4 inches. Don't confuse more digits after the decimal point or higher resolution with greater accuracy.

Operating Temperature

All electronic components and equipment are designed to operate within a specific temperature range. At temperature extremes, LCD panels pop or crack, batteries leak or become inoperative, and microprocessors lock up. The operating temperature range is particularly important in the sonic measurer because the speed of sound in air is dependent on the temperature. As you'll see later, the unit includes circuitry specifically intended to monitor the air temperature and adjust the calculations for distance accordingly.

Humidity

Water and electronics generally don't mix well. Most open-air electronics have a limited maximum ambient humidity rating, above which problems with arcing, leakage, and corrosion can occur. Most electronics can't tolerate condensing humidity—the equivalent of dew on components and circuit traces—for obvious reasons. On the other extreme, very low humidity presents a problem because electrostatic discharge occurs more readily.

In addition, as with temperature, the level of humidity affects the propagation of sound in the air. This inexpensive sonic measurer has no provision for monitoring and adjusting the calculation of distance to account for variations in humidity, however.

Aperture Angle

As discussed, the aperture angle of the transmit/receive horn affects the range and accuracy of the measurements. The narrower the aperture angle, the greater the potential maximum range, assuming a perfectly perpendicular target. The larger the aperture angle, the more likely an echo will be recovered, as well as reflections and other noise that could result in erroneous readings.

Laser Specifications

The Code of Federal Regulations (CFR) Title 21, Part 1040, establishes the safety and performance standards for light-emitting products. Ratings range from Class I (no hazard) to Class IV (an acute hazard to the skin and eyes from direct or scattered radiation). This sonic measurer employs a Class IIIa laser, which is considered an acute viewing hazard. Output of the 630–670nm orange-red laser is limited to 5mW. The increasingly popular green lasers, in comparison, operate between 520 and 565nm, and common infrared lasers generally operate at a wavelength of greater than 800nm.

Don't look directly at the beam or its reflection in a mirror. Your retinas can be permanently scarred.

Operation

Operating the sonic measurer is a simple matter of depressing the soft elastomeric PWR (power) button, securing the unit against a wall or other fixed position, and pressing the READ button. The unit first energizes the laser for a couple seconds and then transmits a few rapid pulses of ultrasound or chirps. If all is well, the reading, in either feet or meters, is displayed on the LCD about a second after the chirps have ceased. The distance measurement is made from the piezoelectric transducer at the back of the unit to the wall or other large reflective target.

If the wall or other object within range isn't flat, isn't perpendicular to the sonic measurer, isn't fully covered by the laser pattern (which approximates a ±5-degree cone), or significantly absorbs or disperses the energy of the ultrasonic chirps, the unit will display an error message, "ERR." If the target is within range and is a flat, reflective wall, perpendicular to the unit, the sonic measurer will transmit two chirps and then display the distance. Otherwise, the unit will transmit up to about a dozen chirps before returning either the distance measure or an error message.

In addition to taking basic distance measurements, the unit supports calculations that are accessed through the area, volume, double segment, and Recall buttons. The area and double segment functions each take two consecutive measures, A and B, and compute the surface area (A × B) and total distance (A + B), respectively. The volume function takes the result of three consecutive measures, A, B, and C, and computes the volume (A × B × C). Pushing the Recall button lets you step back through the values used for each calculation and repeat a measurement. For example, after measuring the volume of a room, you can review the three measures and repeat the second measure (B). The unit will recalculate and display the volume.

Pressing the PWR button for a second time turns off the unit. The unit will also shut down automatically after 30 seconds of nonactivity.

Teardown

The teardown, illustrated in Figure 11-2, requires about 5 minutes, including extraction of the ultrasound transducer.

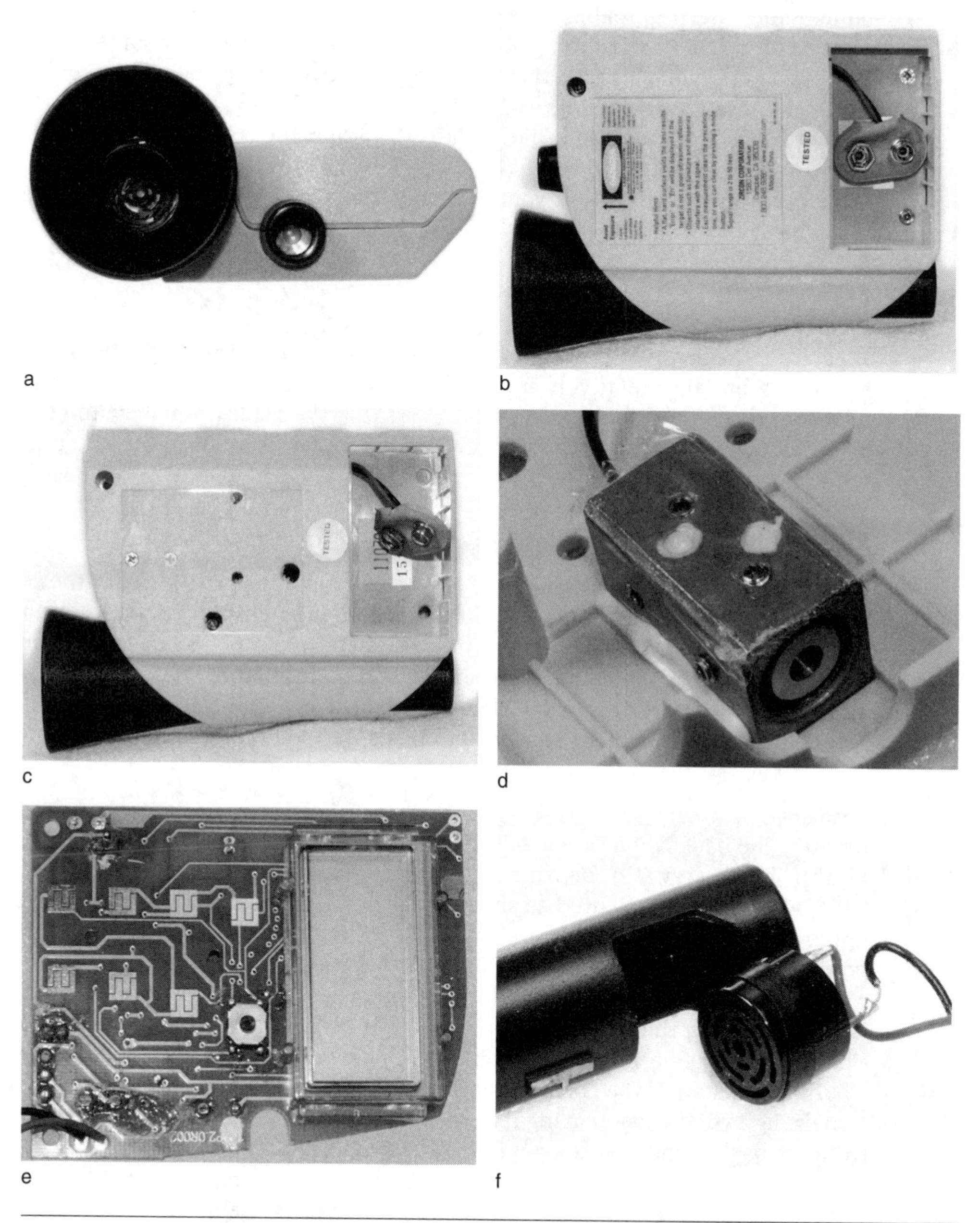

FIGURE 11-2 Teardown sequence

Tools and Instruments

You'll need a sharp knife and a small Phillips-head screwdriver to crack the case. You'll also need an illuminated magnifier and a multimeter with sharp probes to trace the circuitry. The circuitry isn't dense, but the markings on many of the SMT components are impossible to read with the naked eye. You may also need a rough cotton cloth and some rubbing alcohol to remove the glaze over some of the active devices so that you can read their markings. In particular, the coating over the 14-pin IC in my unit was so thick that I couldn't make out a single marking. You may also need a forceps to remove glue between some components and the board.

If your goal is to harvest the SMT components, you'll need a fine-tipped soldering iron and forceps. If you have one, a hot air pen will make the harvest much easier.

Step by Step

If you're following along at home with a sonic measurer, the only thing you have to watch out for is the relatively fragile thermistor bead that's attached to the plastic ultrasound horn. I discovered the hard way that the leads are wispy and easily broken. If you do break a lead, however, don't worry. A bit of solder and tape and it'll be as good as new.

Step 1

Identify the external components. Before you crack the case, identify the ultrasonic transducer in the base of the horn and the emitter of the laser module, which is flush with the top of the unit, as shown in Figure 11-2a. Next, examine the junction of the horn and the case, at the top of the unit. You should see a small, amber bead, about the size of a grain of rice, just inside the seam. This is the epoxy-encapsulated thermistor used to measure the ambient air temperature.

Step 2

Remove the back cover. Remove the battery cover and battery to expose the two Phillips-head screws in the base of the battery compartment. Remove the screws. Next, with the unit oriented as shown in Figure 11-2b, remove the screw in the upper-left corner of the yellow plastic case. If you try to crack the case at this point, you'll encounter stiff resistance.

There are two approaches to dealing with a case that won't open easily. The first is to assume that there's a hidden hinge, clasp, or glue, and apply force until it gives. The second is to search for screws hidden under labels, decals, and rubber feet.

In this instance, the correct approach is to search for hidden screws. To get at the remaining two screws securing the cover, use a sharp knife to peel away the label from the back of the unit, shown in Figure 11-2c. The screws holding the back cover in place are along the top and bottom edges of the newly exposed area. The pair of screws to the left of midline secure the laser module to the inside of the back cover.

Step 3

Free the battery connector and thermistor. As you remove the back cover, note the position of the thin white wires from the circuit board to the thermistor bead identified earlier. Next, push the 9V battery connector through the slot in the plastic back and allow the case to hinge open. Now, carefully remove the thermistor from the horn by tugging on the glue with needle-nose pliers. Avoid the wires and the epoxy bead, which are easily damaged. The plastic back is still tethered by the laser module.

Step 4

Extract the laser module. Free the laser module, shown in Figure 11-2d. The module is secured to the back cover by the two screws uncovered in step 2.

Step 5

Extract the circuit board. Remove the two Phillips-head screws at the base of the circuit board. Don't touch the four small screws that attach the LCD module to the board, near the black epoxy blob. Flip the board over and remove the one-piece set of elastomeric buttons, as in Figure 11-2e.

Step 6 (optional)

Extract the ultrasonic transducer. You'll need an intact horn-transducer assembly to test the circuitry, but if you want a good look at the transducer, simply pop it out of the horn housing, as shown in Figure 11-2f. The transducer is held in place by a friction-fit that's sealed with a black rubber band around the transducer. Push back, toward the lead, to free the cylindrical transducer, which looks like an ordinary piezoelectric buzzer.

Layout

Figure 11-3 shows the layout of the approximately 3 × 2 inch circuit board from the back or component side, oriented with the base of the unit to the left. There are four overlapping areas of the double-sided board that roughly correspond to the functions of power regulation, laser drive, ultrasonic transmission and reception, and numeric processing and display.

In tracing these circuits, I found a short piece of 32AWG wire invaluable. Many of the traces jump from one side of the board to the other, through small vias. When you find a trace that suddenly ends at a plated through-hole or via, insert the wire into the via and flip the board over and continue following the trace marked by the wire.

The power regulation and switching circuitry is in the lower-left corner, where the leads from the 9V battery attach to the board. Key components include a 10μf at 16VDC electrolytic capacitor, a glass SMT diode (D2) that protects against a

FIGURE 11-3 Circuit board layout, component side

reversed polarity battery, a 3-pin 7130 voltage regulator (U4) that supplies 3.0VDC continuously, and an 8-pin LM317LM voltage regulator (U3) that delivers 3.0VDC when the sonic measurer is powered up.

The laser driver circuitry is in the lower-right corner of the board, where the ultrasonic transducer leads attach to the board. There are two 1F NPN and two 5C PNP transistors in the area. One 5C (Q7) and one 1F (Q5) are driven by the microcontroller to switch the laser diode on and off. If you follow the traces, you'll find that the pair of transistors is supplied with 3.0VDC from the LM317LM voltage regulator. The 1F transistor is driven by the microcontroller, through a 310Ω resistor (R33). The other transistor pair in the area controls the distribution of 9VDC to components, including the LM317LM regulator.

The ultrasonic transceiver circuitry is concentrated in the upper-left quadrant of the board. The most prominent components are the 14-pin 274C quad operational amplifier (U2) that serves as the ultrasound receiver, a pair of zener diodes (ZD1 and ZD2) across the leads to the transducer, a 220µf at 16VDC electrolytic capacitor, the output transformer (T1), an inductor (L1), and a variable capacitor (VC1). The variable capacitor, inductor, and zener diodes in my unit were covered in semitransparent glue that required about 5 minutes of work with a forceps to remove. There are also a pair of 1F NPN transistors, a 6C NPN transistor, a 3F PNP transistor, and a pair of glass diodes (D1 and D3) in the area. The 4.1943MHz crystal (X2), located just above the 274C IC, is connected directly to the microcontroller.

The microcontroller and LCD driver circuitry is roughly confined to the right of the board. Several devices connect directly to the input or output ports of the entombed microcontroller, including the LCD module that is connected directly to the microcontroller, the crystal (X2), and the thermistor (TH1). The thermistor and calibration potentiometer (P1) connect to the microcontroller at the bottom edge of the board, between the LM317LM voltage regulator and microcontroller.

The front of the board is relatively sparse, as shown in Figure 11-2e. In addition to the LCD module is a momentary push button and pads for seven elastomeric buttons. No components are hidden under the LCD module. If you've followed the previous teardowns in this book that feature LCD modules, there's no need for you to remove the module to peek underneath. Besides, we'll need the display to test the other circuitry on the board.

Components

Now that you're oriented to the broad functionality of regions on the board, let's take a look at the individual components. Remember that the best resource for component information is manufacturer datasheets, available from the manufacturer's web site or the catalog of a large parts retailer, such as Digi-Key (www.digikey.com) or Mouser (www.mouser.com).

Ultrasonic Transducer and Horn

The 5-inch plastic horn serves as both transmitter and receiver antenna. The modest taper of the outside diameter, from 20 to 40mm, constrains the transmitted and received aperture angle to ±5 degrees from center. A close-up of the horn and transducer is shown in Figure 11-2f. The piezoelectric element of the transducer is visible through the cutouts in the front of the component. The resistance of the transducer is greater than 10MΩ, which is typical for a piezo transducer.

Thermistor

The 1mm diameter thermistor bead and relatively fragile wire leads, shown in Figure 11-4, provide the basis for adjustments in distance calculations based on ambient temperature. Thermistors are manufactured with metal oxides in a way that enhances their sensitivity to change in ambient temperature. This thermistor contains predominantly iron oxide—you can verify this with a magnet.

The ideal thermistor presents a rapid, significant, linear change in resistance with relatively small changes in ambient temperature. This contrasts with typical metal film resistors, which are formulated to present a constant resistance over a wide range of ambient temperatures.

This thermistor has a negative coefficient, meaning temperature and resistance are inversely correlated. Increases in ambient temperature result in decreased thermistor resistance, and vice versa. At an ambient temperature of 79°F, I measured the resistance as 76K. When I raised the temperature of the thermistor to 95°F by holding the bead between my fingers for 5 seconds, resistance dropped to 52K. I used a Fluke 80T-150U temperature probe and 87 DMM to measure the temperature of the air and of my skin.

FIGURE 11-4 Thermistor bead

The thermistor is wired in series with a 9KΩ SMT potentiometer (P2) that can be adjusted to calibrate the unit. The two components are connected directly to the microcontroller.

See the teardown of the hygro thermometer in Chapter 8 for another example of a thermistor circuit.

Laser Module

The laser is the most entertaining find in this teardown. At 1.2 ounces, the laser module is also the heaviest component in the unit. The cylindrical laser diode and lens assembly are embedded in an aluminum block that's about 1-inch long and a little more than 1/2-inch square, shown in Figure 11-2d.

When the laser module is activated with 2.9VDC, it draws 28mA, or about 93mW ($P = 2.9V \times 0.028mA$). Assuming a typical efficiency of 1 percent, that's less than 1mW output. The majority of power is dissipated as heat, which explains the relatively massive block of aluminum—it serves as a heat sink. Laser diode efficiency and lifespan both suffer at elevated temperatures.

LCD Module

The LCD module in this teardown is essentially identical to the LCDs encountered in teardowns in earlier chapters. For example, as with the bathroom scale and pedometer (Chapters 3 and 5), the LCD panel interfaces with traces on the circuit board through an elastic Zebra connector.

Microcontroller

As noted in previous teardowns that featured epoxy-entombed microcontrollers, including the bathroom scale and pedometer, I consider the hidden slab of silicon a microcontroller instead of a microprocessor because of onboard analog-to-digital (A–D) processing of sensor data.

This microcontroller controls the LCD display, ultrasonic transceiver, and laser module. It also reads and interprets the thermistor value and performs calculations for distance, surface area, and volume. Its internal clock or timebase is dependent on the accuracy and stability of the 4.1943MHz crystal (X2).

7130 Positive Fixed Voltage Regulator

The 7130 fixed voltage regulator (U4) is the 3-pin SOT-89 component just to the left of the electrolytic capacitor, shown in Figure 11-5. The device receives input voltage from the 9VDC battery, via glass diode D2. The tab and middle pin are input, the top pin is ground, and the bottom pin is output. According to the datasheet, the regulator accepts an input of up to 24VDC and provides an output of 3.0VDC at 30mA. Maximum no-load current is 6μA.

The no-load current figure is important in this application because, as you can see in the figure, there is no on–off switch between the battery input and the regulator. I measured the quiescent current flow to the regulator with a fresh 9V battery at 2μA.

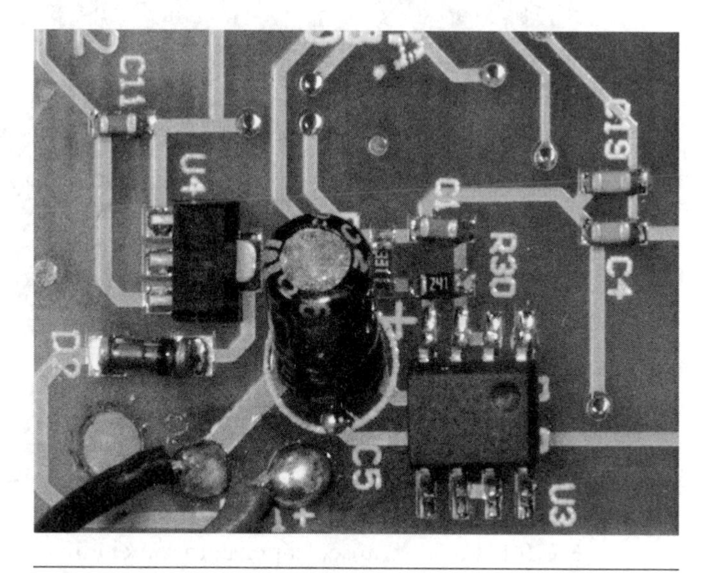

FIGURE 11-5 Voltage regulator section with 3-pin 7130 and 8-pin LM317LM

LM317LM Positive Adjustable Voltage Regulator

The LM317LM voltage regulator (U3) is the SOP-8 component visible in Figure 11-5. Unlike the smaller 7130, which can output only 3.0VDC, this regulator can be configured with external resistors to provide between 1.2 and 28VDC at 100mA. In this application, the output is fixed at 3.0VDC at 100mA.

The LM317LM, which is internally protected against shorts and elevated temperatures, is switched on when the PWR switch is depressed and switched off if there is no activity for 30 seconds. Output is on pins 2, 3, 6, and 7. Input is pin 1, and pins 5 and 8 are for voltage adjustment resistors.

ST274C Quad CMOS Operational Amplifier

The ST24C, which contains four independent CMOS (complementary metal oxide semiconductor) operational amplifiers, is the SO-14 package visible in Figure 11-6. Relevant specifications include low power consumption, owing to the CMOS construction; a supply voltage requirement of from 3 to 16VDC; and a gain bandwidth product (GBP) of 3.5MHz. The GPB is the midband gain of an operational amplifier multiplied by its bandwidth in MHz. In other words, amplifying a 40kHz signal is easily within the capability of the chip.

As noted in the teardown, a semitransparent film obscures the package markings. In this figure, pin 1 is in the upper-right corner of the device. For more information on the ST274C, consult the STMicroelectronics web site (www.st.com).

FIGURE 11-6 Transceiver circuitry, including 14-pin ST274C

Transistors

Four models of discrete SMT transistors are on the circuit board, all in SOT-23 packages. This packaging makes circuit tracing a breeze. Looking at the 3-pin transistor package from above, with the single pin up and two pins down, the collector is on top, the base on the lower-left, and emitter on the lower-right.

The 1F (BC847B) and 6C (BC817-40) transistors are general-purpose silicon NPN transistors rated at 50V. The 1F transistor is rated at 100mA collector current and the 6C transistor at 500mA. The 3F (BC857B) and 5C (BC807-40) general-purpose PNP transistors are rated at 45V, with a maximum collector current of 100 and 500mA, respectively. These transistors are employed in NPN-PNP pairs, with the transistor with the lower current rating driving the larger capacity transistor. The transistors listed in parentheses are common equivalents.

Diodes

A few unmarked, apparently general-purpose, glass diodes are also on the board. In addition, two unmarked SMT zener diodes are across the leads of the ultrasound transducer. Recall that zener diodes break down gracefully and exhibit a constant voltage drop, as long as the reverse current doesn't exceed the capacity of the diode. In this circuit, the diodes are used as a voltage clamp to protect the ultrasonic transducer from damage due to overvoltage. Based on my measurements, these are 75V zener diodes configured to constrain the output to the ultrasonic transducer to 150V peak-to-peak.

Crystal

The 4.1943MHz quartz crystal, visible in the top-right corner of Figure 11-6, is central to the operation of the microcontroller. The route from crystal to microcontroller, near test point 2 (TP2), is circuitous, involving traces on both sides of the board, including a trace under the LCD module. The crystal serves as the timebase for the microcontroller. This is a critical function, given that microseconds matter in measuring the speed of sound.

Quartz crystals are rated in terms of frequency tolerance in parts per million (ppm), operating temperature range, load and shunt capacitance, drive level, and ageing in ppm. Load and shunt capacitance and drive level affect component selection during circuit design. Most important from the perspective of sonic measurement accuracy are the frequency tolerance, operating temperature range, and ageing. Frequency tolerance, like resistor or capacitor tolerance, denotes the accuracy of the component at the time of purchase. Crystal ageing, on the other hand, indicates the likely change in resonant frequency while used in a circuit. Heat, vibration, stress from applied voltages, and physical stress from the internal mount all influence how a crystal will age.

Assuming X1 is an inexpensive quartz crystal, it probably has frequency tolerance of ±50ppm, or ±0.005 percent, and ages at ±5ppm, or ±0.0005 percent, per year. According to these specifications, the crystal should oscillate at 4.1943MHz ±210Hz and age ±21Hz per year. These values are overly optimistic, however, because they assume a constant operating temperature, which is an unrealistic assumption with a lightweight, handheld instrument. Precision measurement instruments designed for professional work often place quartz crystals used as a timebase in an insulated chamber that's maintained at a constant, elevated temperature.

Inductors

The two inductors on the board, the transformer (T1) and resonator (L1), are in the ultrasonic transceiver section, as shown in Figure 11-6. Figure 11-7 shows the transformer with its insulation partially removed to reveal the secondary winding composed of 46AWG enameled solid copper wire. The leads to the primary winding of T1 exit the top of the component. The secondary winding is attached to pins on the bottom of the transformer. Both inductors are wound on ferrite cores. You should verify core composition with a magnet. Many cores that look like ferrite are actually plastic.

I measured the inductance of L1 as 28mH at 1kHz and through a destructive teardown found that it's composed of more than 100 turns of 46AWG enameled

FIGURE 11-7 Transformer with insulation partially removed

copper wire wound tightly over the ferrite core. This inductor is used as part of a filter in the receiver circuit.

The secondary of transformer T1, like L1, is composed of at least 100 turns of fine enameled copper wire that's apparently 46AWG. The primary, which exits at the top of the component, is composed of about 40 turns of 32AWG solid enameled copper wire. The DC resistance of the secondary winding is 80Ω and that of the primary winding is 0.1Ω.

Switches

Seven of the eight switches on the front of the unit are elastomeric. As detailed in the teardown of the electronic pedometer in Chapter 5, depressing a carbon-impregnated elastomeric button shorts underlying traces on the circuit board, completing the circuit. The exception is the READ switch, which is a typical momentary-on SMT switch that is soldered directly over the board tracing of an elastomeric switch. That is, the mechanical switch is an afterthought, a modification made after the board was released.

One reason for using the mechanical switch may be because it's the most often used switch and more prone to failure. This supposes a greater MTBF (mean time before failure) for the particular mechanical switch versus the elastomeric switch. Can you think of another reason for installing the mechanical switch over the existing elastomeric switch?

Capacitors

The most notable capacitor on the board is the ceramic trimmer capacitor in the transceiver circuit, clearly visible in the upper-left corner of Figure 11-6. Trimmer or variable capacitors are considerably more expensive than fixed capacitors of the same voltage rating, but they allow fine-tuning of analog circuits. This capacitor is part of the LC filter in the receiver.

There are also two leaded electrolytic capacitors, and about a dozen garden-variety ceramic SMT capacitors. The 10µf at 16V electrolytic capacitor is used on the output of the LM317LM 3.0V regulator. The second electrolytic, a 220µf at 16V capacitor, is used on the 9V line feeding the switching transistors in the transmitter circuit.

Resistors

The board is populated with standard-tolerance SMT resistors. By now, you should be able to determine the value of resistors, including those with letter markings. If not, check out Appendix A to review SMT resistor markings.

How It Works

Because of the complexity of the physics involved in the operation of the sonic measurer, I've broken down the discussion into operational and functional perspectives.

Operational View

Let's start with an operational view of the sonic measurer, using Figure 11-8 as a guide. As shown in the top half of the figure, a transducer horn transmits a series of 40kHz pulse trains or chirps, separated by 110ms. Each chirp is composed of perhaps a dozen 40kHz pulses. After a little over 25ms, the ultrasonic wave train hits the interface between the air and the target—in this case, a perfectly perpendicular, reflective wall that extends infinitely in all directions.

At the interface, some percentage of the wave train—the echo—is reflected back toward the horn. The part of the echo that is captured by the horn causes the piezoelectric transducer to vibrate, generating an electric pulse. The microcontroller determines the time between events and stores the number. This process is repeated for the second chirp.

Now let's examine the amplitude versus time relationships in the lower half of Figure 11-8. Notice how the recovered pulse or echo is depicted as lower in amplitude and less sharply defined relative to the original pulse train or chirp.

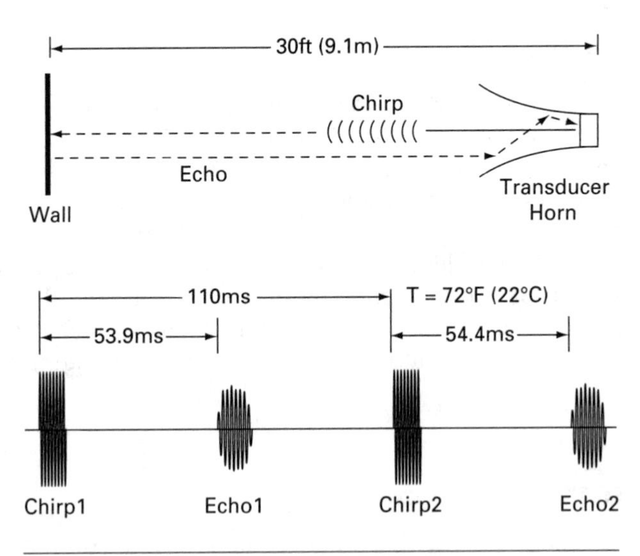

FIGURE 11-8 Operational view of the sonic measurer

This is because the energy of the original chirp is dispersed and attenuated by the air as it leaves the horn and is partially absorbed by the target 30 feet away. The ultrasonic chirp is additionally dispersed and attenuated on the return trip to the horn. If drawn to scale, the echo would not be visible on the amplitude tracing in Figure 11-8.

The interval between the start of each chirp and the start of its corresponding echo is computed by the microcontroller. In this example, the first measurement returned by the sonic measurer is 53.9ms. The interval between the leading edge of the second chirp and that of its echo is longer, at 54.4ms. The microcontroller compares the two interval values and, if they're close enough in value, a distance measure is flashed on the LCD screen. That measure may be the distance computed from the first interval, the second interval, or the average of the two intervals. My vote is for the average.

If the two interval values are too dissimilar, then the microcontroller will either transmit and evaluate another pair of chirps and echoes or generate an error message on the LCD. In my experimentation with the unit, I was unable to get it to repeat the chirp repetition cycle more than six times before it presented an error message. I assume that the similarity score is computed as a percentage of total time. Otherwise, longer intervals would require increased accuracy to get through the system without an error message. You'd need access to the microcontroller source code to verify the algorithms.

Now let's consider the physics of ultrasound in air. The nominal speed of sound in air, 1130 feet per second (ft/s), is an approximation for dry air at room temperature, at sea level, and with a typical CO_2 concentration. A more accurate figure for the velocity of sound in meters per second (m/s) considers the temperature, humidity, and CO_2 concentration, in the form of the following equation:

$$\text{Velocity of Sound} = (n\text{RT/M})^{1/2} \text{ m/s}$$

where n is the adiabatic constant, a characteristic of the gases in air (nominally 1.4), R is the universal gas constant (8.314J/mol K), T is the absolute temperature in degrees Kelvin, and M is the molecular weight of the gases in kg/mol. According to the equation, the speed of sound in air is proportional to the square root of the absolute temperature and increases slightly with increasing humidity and CO_2 concentration. In practical terms, the speed of sound in air is a function of the density of air, which is dependent on the temperature.

Solving the equation with reasonable speed and accuracy is problematic using integer arithmetic on a typical microcontroller, and I doubt that the chip in our sonic measurer is capable of floating-point math. A more microcontroller-friendly model for the velocity of sound in air, and the one probably used in the sonic measurer, is this:

$$\text{Velocity of Sound} = 331.3 + 0.6\text{C m/s}$$

The 331.3 figure is the speed of sound in m/sec at 0°C, assuming 0 percent humidity. This approximation ignores the contributions of humidity, CO_2, and the frequency of the sound on velocity.

Let's return to Figure 11-8, where the back of the transducer horn is 30 feet from the wall and the ambient temperature is 72°F. According to the second formula, the velocity of sound is 345.5m/s (332.3 + 0.6 × 22) or 1113.5 feet/s. The round-trip transit time for an ultrasonic chirp in this environment is (30 + 30) / 1113.5, or 53.9ms. The first chirp-echo interval agrees with this theoretical figure. The second chirp-echo interval is longer than expected.

Using the same approach, within the operating temperature range of the sonic measurer, 20–120°F, transit time ranges from 55.7ms at 20°F to 50.5ms at 120°F. This 5.2ms difference corresponds to distance of nearly 6 feet at 72°F. Obviously, the sonic measurer has to control for temperature to provide distance measurements within the specified accuracy.

Functional View

Now consider a functional view of the sonic measurer, using the simplified schematic of the systems in Figure 11-9 as a reference. Let's focus first on the power circuitry, which is shown in the upper-left of the schematic.

A 9V alkaline battery supplies power continuously. There is no hard on–off switch. Based on my measurements, the current demanded of the battery varies

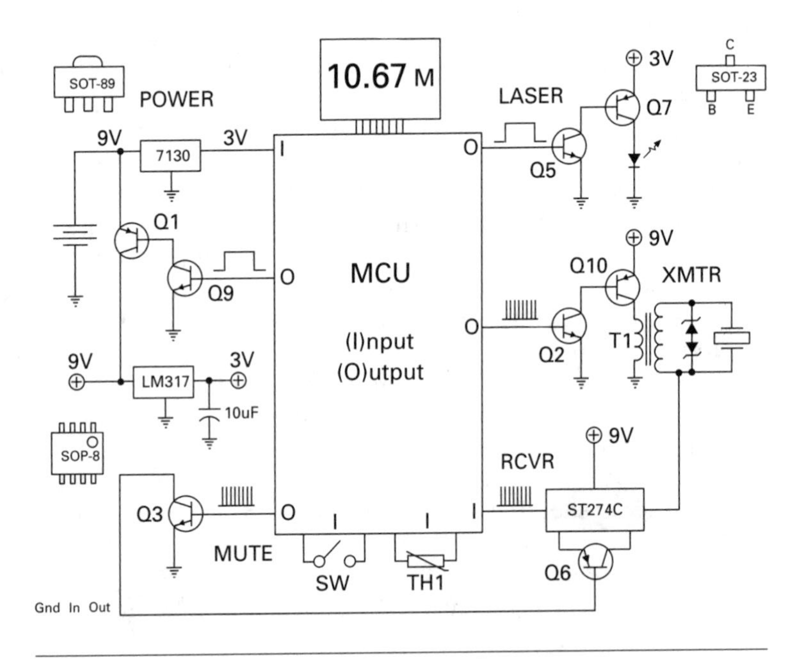

FIGURE 11-9 Simplified schematic of major systems

from 2μA when the unit is hibernating to 40mA when the laser is on and ultrasonic transmitter is active. Assuming a standard 300mAh (milliamp hours) alkaline battery, the specification of 8-hour battery life with continuous use seems reasonable (300mAh/40mA = 7.5 hours).

Three different supply lines power the circuitry in the sonic measurer. The 7130 voltage regulator provides the microcontroller with 3.0VDC at all times. The switches on the front of the circuit board and the thermistor, which interface directly to the microcontroller, are also continually energized by the output of the 7130 regulator.

The second 3.0VDC line, supplied by the output of the LM317LM, is live only when the 9VDC input to the LM317LM is switched on by the microcontroller through the 1F (Q9) NPN and 5C (Q1) PNP transistor pair in the laser area of the board. When the output of the microcontroller is high, Q9 is forward-biased, which in turn forward-biases Q1, providing a low resistance path from the 9V battery to the input of the LM317LM. In addition to powering the LM317LM, the switched 9VDC also supplies the transceiver circuitry, including the ST274C quad op amp (U2) and associated transistors Q2, Q10, Q3, and Q6.

The basic switching configuration used with Q1 and Q9 is repeated in the laser circuit and twice in the transceiver circuit. In the laser circuit, shown in the upper-right corner of the schematic, the output of the microcontroller goes high, forward-biasing Q5, enabling Q7 and the laser diode to conduct. The voltage across the laser diode is a few tenths of a volt less than the voltage supplied by the LM317LM voltage regulator because of the voltage drop across the junctions of Q7. I didn't crack the laser diode case, but I assume the module contains a series resistor to limit current flow through the diode.

The microcontroller is the source of the 40kHz signal that's eventually transmitted. The transistor pair Q2 and Q10 and step-up transformer T1 are used to increase the amplitude of the signal to achieve a range of at least 50 feet. A signal train of 40kHz pulses is output from the microcontroller, which forward-biases Q2 and Q10, sending pulses from the switched 9VDC source through the primary of transformer T1 to ground. These 9VDC pulses at 40kHz are stepped up by T1 to 150VPP (peak-to-peak). That is, T1 must have a turns ratio of at least 150/9, or 16:1. A pair of 75V zener diodes is placed back-to-back across the transformer secondary to limit voltage across the piezoelectric transducer.

The microcontroller, which is in hibernation mode until the PWR switch (SW) is depressed, employs its own internal power switching circuitry. In addition to monitoring the soft switches, the microcontroller maintains voltage across the thermistor at all times. This minimizes fluctuations in the temperature of the thermistor that would otherwise occur due to self-heating if the thermistor had to be energized before each reading.

As discussed in the teardown of the hygro thermometer in Chapter 8, the process of measuring the resistance of a thermistor involves applying either a continuous or pulsed voltage across the thermistor. The resultant current heats the thermistor, raising its temperature above the ambient temperature. Depending on ambient conditions, this self-heating could be a source of error.

To give you an idea of the significance of the thermistor, consider that I measured a distance with the sonic measurer as 4 feet, 11 inches at ambient temperature of 79°F. I then held the thermistor bead between my fingers, raising the temperature of the bead to 95°F. The instrument read the same distance as 5 feet, 0 inches. That's a difference of 1 inch at only 5 feet. The difference would be proportionally greater at greater distances.

As in most transceivers, the receiver circuitry is considerably more complex than that of the transmitter. Whereas the transmitter circuitry simply has to amplify the pulse train from the microcontroller, the receiver has to distinguish the incoming signals from noise and ignore the transmitted pulses.

As show in simplified form in the lower-right area of Figure 11-9, the analog signal processing for the receive function is performed by the ST274C quad op amp, using the signal produced by the piezoelectric transducer when it receives an ultrasonic impulse. A muting circuit consists of a 3F (Q6) and 1F (Q3) transistor configured in the familiar pairing used with the other systems in the unit.

A more detailed, but still simplified, schematic of the receiver circuit is shown in Figure 11-10. For clarity, I haven't included many of the bypass and feedback capacitors and resistors. I also don't show the zener diodes or transformer secondary across the ultrasonic transducer. Note the two analog data feeds to the microcontroller, from op amps 1 and 3, and that the muting function is controlled by a signal from the microcontroller.

Before continuing with the circuit, let's revisit the needs of our receiver. The first requirement is that it receives echoes from the transmitted chirps or ultrasonic

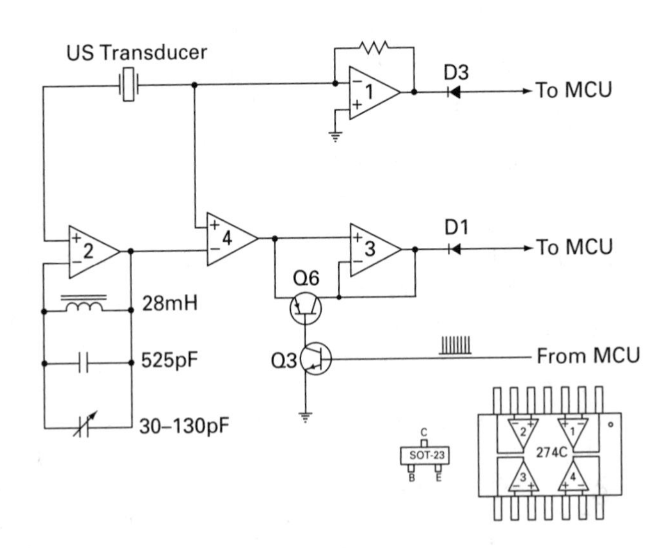

FIGURE 11-10 Simplified schematic of the receiver

pulse trains. The second requirement is that it ignores noise in the environment—that is, the receiver should be selective and provide a good signal-to-noise ratio. If you consider the purpose of the sonic measurer, it's likely to be used in an environment with noise from power tools, ultrasonic rodent repellers, and other noise sources that extend to 40kHz and beyond. The plastic horn provides a degree of directional selectivity, which cuts down on noise. The ultrasonic transducer also has a natural resonant frequency, presumably around 40kHz, that provides attenuation to signals greater and less than 40kHz.

Now, referring back to the schematic in Figure 11-10, let's assume that the piezoelectric transducer is generating an electrical signal secondary to ultrasonic vibrations—that is, there's an AC signal across the transducer.

Op amp 1 is configured as a basic inverting amplifier, meaning that it takes the signal at the inverting (–) input and amplifies it, with the output 180 degrees out of phase with the input. (By the way, if you're new to op amps (operational amplifiers), they're covered in more detail in Chapter 13, a teardown of a guitar effects pedal.) For now, let's stay with a relatively high-level discussion. The output is sent, through diode D3, to the microcontroller (MCU). Because the diode has a minimum forward voltage drop of about 0.7V, only signals greater than that make it to the input of the microcontroller. Noise and weak echoes are blocked by the diode.

In a quiet, indoor setting, a receiver composed of op amp 1 and D3 would probably be sufficient. However, other than the selectivity provided by the transducer and horn assembly, there is no provision for noise rejection by the op amp 1 circuit. Any sound impacting the transducer will result in a signal to the microcontroller.

The remaining three op amps in the 274C chip are used to increase the effective signal-to-noise ratio of the receiver. Op amp 2 is configured as a noninverting amplifier with unity gain at 40kHz—that is, a notch filter. At resonance, the feedback network composed of a parallel resonant LC circuit from the output to the inverting input (–) of op amp 2 is essentially a wire. Above and below 40kHz, the impedance of the feedback increases, thereby increasing the gain of the amplifier circuit.

Recall the resonant frequency of an LC circuit is $1 / (2\pi(LC)^{1/2})$. Applying the formula with a 28mH inductor (L1) in parallel with a 525pF fixed capacitance and a 30–130pF variable capacitance results in a resonant frequency range of about 37 to 40kHz. The variable capacitor enables the manufacturer to tune the circuit to precisely 40kHz. The alternative, using an inductor and fixed capacitor with high tolerance, was probably cost-prohibitive.

Op amp 4 compares the unfiltered signal on one side of the transducer with the output of op amp 2, taken from the other side of the transducer. The input signals to the two op amps are thus 180 degrees out of phase. If the signal from the transducer is 40kHz, then the amplitude of the signals on the inverting (–) and noninverting (+) inputs to op amp 4 should be equal. Assume that at one instant the input to op amp 4 is positive and the input to op amp 2 is negative. The positive input to the noninverting (+) input of op amp 4 results in a positive output. Similarly, the negative input to the inverting (–) input of op amp 4 results in a positive output.

Now let's assume a signal of broadband noise source that extends from 30 to 50kHz. Op amp 1 will amplify the signal and send it to the microcontroller. Op amp 2 will amplify it with a gain of greater than unity. Now op amp 4 is presented with a strong signal on its inverting (–) input, relative to the direct signal from the transducer on its noninverting (+) input. The result is an increase in the amplitude in the output of op amp 4, relative to the output with a 40kHz signal.

Ignoring the contribution of the Q3-Q6 transistor pair for the moment, op amp 3, which is configured as a unity gain buffer, takes the output of op amp 4 and sends it, through diode D1, to the microcontroller. D1 serves the same noise-limiting function as D3. In summary, at this point in our analysis, when a pure 40kHz sound is impacting the transducer, the output from op amps 1 and 3 are about the same in amplitude. When a noise source extends on either side of 40kHz, the output of op amp 3 may be many times that of op amp 1.

My assumption is that the program running on the microcontroller examines the relative amplitudes of both signals to determine whether to return an error message or use the signal to compute distance. As long as the amplitudes of the signals from op amps 1 and 3 are about the same amplitude and great enough to overcome the noise barriers provided by D1 and D3, the signal is considered a valid echo. When the amplitude of the output of op amp 3 is significantly greater than that of op amp 1, the signal is presumably considered noise by the microcontroller.

Finally, let's consider the muting circuit formed by Q6 and Q3. As you can see in the schematic, the collector and emitter of Q6 are across the input ports of op amp 3. As a result, when Q6 conducts, the input ports are shorted and the output of Q3 is zero, regardless of the output of op amp 4. Q3 receives 40kHz pulses from the microcontroller that are identical to those sent to the transmitter. As a result, the output of op amp 3 is muted during transmit.

Given that the microcontroller can be programmed to ignore signals that occur during transmit and that the output of op amp 1 is not muted, my assumption is that the output of op amp 3, if not muted, could damage the input circuitry of the microcontroller. To verify this assumption, we'd have to have access to information on the entombed microcontroller. Can you think of another reason for the muting circuit? Why not simply limit the input to the microcontroller with a diode clamping circuit?

Mods

This sonic measurer is more of a platform for experimentation and repurposing than for mods. The laser module with its built-in ±5-degrees targeting cone pattern is an interesting tool in itself. Given the spread of the pattern, it's easy to estimate the distance of a moving target or of a structure more than 50 feet away. Recall that the ultrasonic measurer requires a stationary target.

Another repurposing project that you might consider is an ultrasonic source detector. Simply remove D3 from the output of op amp 1 and either replace it with an LED or the leads of your multimeter. You might be surprised at the number of significant ultrasound generators in the environment. For example, I discovered that when my bicycle chain needs lubrication, it generates a lot of ultrasonic noise when I turn the pedals with the bike on a stand, even though I can't hear any change in the sound produced by the chain.

I'm sure you can think of ultrasonic diagnostic aids to use with home appliances, car engines, and other mechanical devices. There are also uses for ultrasound as the basis for biological experimentation, from disrupting the flight of bats, to chasing away stray dogs and irritating rodents.

PART III

For Musicians

Chapter 12

Electric Guitar

The Fender Telecaster, or "Tele," the first mass-produced, solid-body electric guitar, is the easiest guitar to tear down and mod. And there's good reason—sheer simplicity. As shown in Figure 12-1, the basic design hasn't changed since Fender introduced the guitar in 1949—a solid body, bolt-on neck, one or two magnetic pickups, a volume control, and what passes for a tone control. Models with two pickups have the added complexity of a pickup selector switch, but a stock Tele has no whammy, or tremolo, bar.

In this chapter, we'll tear down a Fender American Deluxe Telecaster, which incorporates two pickups, a pickup selector switch, and volume and tone controls. I'll also walk you through a mini-teardown of a magnetic pickup and discuss several mods, including how to configure the Tele as a MIDI (Musical Instrument Digital Interface) controller.

Highlights

This teardown is deceptively simple. After all, we're working with a slab of wood, six steel strings, tuning machines, a pair of magnetic pickups and potentiometers, a capacitor, a three-way switch, some wire, and an audio output jack. As electronic devices go, it doesn't get much simpler—or does it? As you'll see, there's a hidden elegance in the simplicity of the Tele, and what appears simple at the outset is actually a fairly complex electromechanical device. During the teardown, note the following:

- How shielding is used to minimize EMI
- A star ground configuration that minimizes noise
- Construction and wiring of the pickups—a major contributor to the guitar's tone
- Reliance on traditional components and wiring

FIGURE 12-1 Fender American Deluxe Telecaster

Specifications

Guitar specifications vary markedly from one manufacturer to the next, and even within the same brand. Consider that Fender Teles retail from about $180 to $1800, depending on the quality and origin of the body, neck, and components. You can pick up a knock-off from dozens of vendors for less than $200, and online suppliers offer everything from replacement parts to complete kits. Because they're so easy to mod, you never really know what's inside a pre-owned Tele until you take a look inside. For example, this teardown features a Tele in which I replaced the stock potentiometer and capacitor with a more functional tone control.

If you've shopped recently for an electric guitar, you know that even when specifications are offered, they tend to be qualitative and focused on materials and construction. There's typically nothing quantitative in the way of output level, signal-to-noise (SNR) ratio, or tuning stability. Instead, the marketing brochures list the type of wood—typically swamp ash or alder—and finish, neck configuration and color, and the type of mechanical tuners. There might be a mention of the types of pickups. For example, here's a condensed version of the specifications for the American Deluxe Ash Telecaster, from the Fender site (www.fender.com):

- Ash body
- U-shaped maple neck
- Fingerboard with vintage tint, abalone dot position inlays, and 22 medium jumbo frets
- Stainless steel Telecaster bridge with chrome-plated solid-brass saddles
- Samarium Cobalt Noiseless Telecaster pickups
- S-1 switching

Although everything listed here affects tone and playability, the types of pickups and switching system are the most relevant specifications for our purposes. "S-1 switching" is simply Fender's marketing term for a four-way switch.

Figure 12-2 shows the key electronic elements of a Telecaster. On the far right along the midline, directly under the six steel strings, is the neck magnetic pickup, which is mounted on a springy suspension system in the swirl-patterned plastic pickguard. The strings terminate in a fixed bridge that's bolted directly to the body. Mounted diagonally in the bridge is the bridge magnetic pickup.

Along the bottom, mounted in the control plate, are, from left to right, the tone control, volume control, and pickup selector switch. Not shown in Figure 12-2 is the 1/4-inch audio out jack, which exits the side of the guitar at about the level of the tone control.

FIGURE 12-2 Electronic elements of the Telecaster

Operation

Mastering the guitar involves patience and years of practice. Designing a guitar that plays correctly is a nontrivial task that requires knowledge of mechanics, electronics, and how to integrate the two. For example, the height of the strings above the pickups significantly affects the spectral content and amplitude of the output signal. Strings that are too close to the magnetic pickups create a distorted overdriven signal, and strings too far from the pickups produce a weak, noisy signal. Similarly, the metal truss rod that runs through the neck must be adjusted so that there is a slight concavity to the fretboard—otherwise, the strings will buzz when struck. But too much concavity adversely affects playability.

In addition, the distance between the two supports for each string—the nut at the head of the guitar and the saddle where each string terminates in the bridge—must be adjusted for correct intonation. Otherwise, the notes played along one part of the neck will be in tune, but notes played at another part of the neck will be out of tune.

Moreover, these electromechanical systems are interdependent. If you modify the concavity of the neck, you affect intonation, the height of the strings above the magnetic pickups, and the guitar tone. As you can see, that simple slab of wood and strings is actually a complex system, and we've only touched on the circuitry.

From an electronics perspective, an electric guitar is one component in a large, interdependent system that minimally includes cables, an effects pedal, an amplifier, and a speaker cabinet. Each element contributes to the tone or sound of the guitar. Without an amp, an electric guitar is useless. Many players wouldn't consider practicing or playing on anything other than a tube-type amp, but others swear by the lightweight, portable, and more affordable solid-state amps with built-in effects.

Back to the operation of the guitar, the player sets everything in motion by strumming one or more steel strings. The vibrating strings interact with the magnetic fields emanating from the neck and bridge pickups, resulting in the generation of electrical signals. Pressing down on a string so that it makes solid contact with a fret on the neck changes the string's frequency of vibration and the frequency of the signal generated by the magnetic pickups. This signal, which may be mixed with signals from other instruments, is routed to the volume and tone control potentiometers, and finally to the audio out jack.

Note the alignment of the six steel strings over the six pole pieces of the bridge pickup, as shown in Figure 12-3. Pole pieces are metal cylinders—typically magnetic—that are embedded vertically in the pickup. The bridge and neck magnetic pickups each produce a distinctive tone, by virtue of their design, adjustment, and relative position. The three-way selector switch enables the player to select the signal from the neck pickup, the bridge pickup, or neck and bridge pickups in parallel. A popular mod, and a standard feature on some Teles, is a four-position switch that allows a fourth option of connecting the neck and bridge magnetic pickups in series.

The volume and tone controls do more or less what their names suggest. The volume control varies the amplitude of the magnetic pickup signals—typically from zero to a few hundred millivolts at an output impedance of a few hundred-thousand

FIGURE 12-3 Bridge close-up

ohms. The stock tone control mainly attenuates the higher frequencies, with the resulting output varying from sparkling to muddy. Because there isn't much need for a muddy tone, it's a common practice to set the tone potentiometer to high and then leave it alone. A popular mod involves replacing the tone potentiometer with a switched bank of capacitors, which results in brighter tone—that is, more treble.

Teardown

You should be able to complete the teardown, illustrated in Figure 12-4, in about an hour. If you're following along at home with your Tele or another electric guitar, remember that electronic components and wiring configuration vary from one guitar to the next, depending on the make, model, year of production, country of manufacture, and, most importantly, the mods performed by previous owner(s).

If you're tearing down a stock Tele, you'll see essentially the same components and wiring described in this teardown. The only significant deviation from a stock guitar is a tone control mod, which I'll discuss in detail later. During the teardown, make note of the wiring and switching arrangement, as these can significantly affect a guitar's tone.

Tools and Instruments

You'll need a Phillips-head screwdriver, wire cutters, and a multimeter. A magnet polarity tester—a small bar magnet or magnetic compass—is useful, but not necessary. Place a towel, carpet remnant, or foam sheet on your workbench to

FIGURE 12-4 Teardown sequence

protect your guitar's finish. A few hand towels or small foam sheets would be helpful as well.

Step by Step

If you plan to repair or refurbish a guitar you're tearing down at home, try to minimize the number of times you remove and replace the wood screws. It's not a good idea to remove and reinsert the wood screws more than a few times during the life of a guitar, because the wood eventually strips. However, for your benefit, I'll remove all the wood screws that provide access to the electronics.

Step 1

Remove the guitar strings. Remove the strings as you would if you were preparing to restring the guitar. Start by releasing the tension on the strings by unwinding each of the six tuning machines. Cut the strings in half with diagonal cutters or, if you have them, guitar string snips. Either remove the string remnants from the bridge and the tuners, or trim them so that they're out of the way.

Step 2

Release the pickguard. Remove the eight Phillips-head wood screws retaining the pickguard.

Step 3

Examine neck pickup. Flip over the pickguard onto a cloth, as shown in Figure 12-4a. Note the magnetic neck pickup mounted in the pickguard and the two wires running from the pickup toward the control plate. The white or yellow wire is signal and the black wire is ground. As you'll see later, the polarity, or, more properly, the phasing, of the magnetic pickups matters.

Step 4

Examine the cavity. Note that the cavity is thickly coated with dark gray or black conductive paint. You should also see a metal tab attached with a wood screw to the cavity and a wire leading from the tab toward the controls.

Step 5

Verify the conductivity of the cavity coating. With your multimeter, check the resistance from the metal tab to any point within the cavity. The resistance should vary from a few hundred to perhaps a few thousand ohms, depending on the space between the probes. The cavity is painted with conductive paint to shield the electronics from external magnetic fields.

Step 6

Release the bridge. Remove the three Phillips-head wood screws retaining the bridge. Depending on the bridge design, you may have to push the center saddle—one of the metal blocks securing a string to the bridge—aside to access the middle screw. See Figure 12-3 for a view of the six saddles and their attachment to the bridge.

Step 7

Examine the bridge. Flip over the bridge onto a soft cloth or foam sheet, as shown in Figure 12-4b. Note the white signal wires and the black ground wire. Note also the copper braid running from the lug attached to the center of the cavity to the surface of the guitar body. The exposed braid, shown in more detail in Figure 12-4c, is normally sandwiched between the guitar body and the bridge to ground the bridge. The grounded bridge plate and painted cavity serve as electromagnetic shielding for the bridge pickup and associated wiring. Set the bridge back in place for now.

Step 8

Extract the control plate assembly. Remove the Phillips-head wood screws on either end of the control plate. Extract the assembly and lay it across the cavity, as shown in Figure 12-4d. From left to right are the tone potentiometer, volume potentiometer, and pickup selector switch. The two controls are 250KΩ audio potentiometers, and the switch is a Fender three-position pickup selector switch.

Step 9

Examine the control plate cavity. Push the control plate to the side, taking care not to scratch the guitar finish. You should see the conductive paint throughout the cavity, as well as a metal solder tab attached to the cavity with a wood screw, as shown in Figure 12-4e.

Step 10

Trace the black ground wires from the tabs within each cavity. All ground wires should terminate on the back of the volume potentiometer, shown in Figure 12-4f. In my guitar, four wires are soldered to the back of the pot—one for the bridge cavity ground, one for the neck cavity ground, one for the control plate cavity ground, and one for the audio output jack ground.

Layout

Figure 12-5 shows the basic wiring layout of the volume and tone controls found in most electric guitars. Leads from one or more pickups (the white wire at top right) terminate at a volume potentiometer (right) and tone potentiometer with

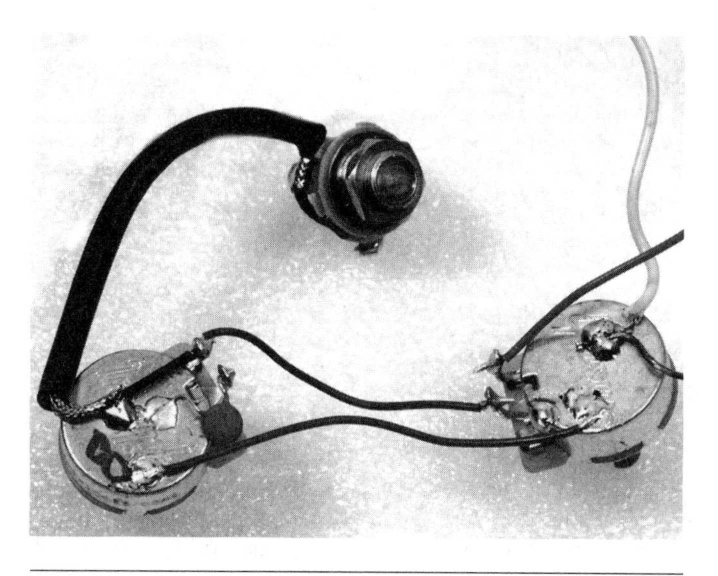

FIGURE 12-5 Basic electric guitar control layout

capacitor (left), and from there, pass to an audio output jack (center). The capacitor and tone potentiometer form a variable low-pass RC (resistor-capacitor) filter. The components shown in Figure 12-5 are from a Gibson Melody Maker guitar, but the same configuration can be found in any number of guitars.

Note that in Figure 12-5, the connection between the tone potentiometer and the output jack is made with coaxial cable. The shield is soldered to the tone potentiometer case, which is in turn wired to the case of the volume potentiometer. A cheaper alternative to coaxial cable is twisted pair, which provides somewhat less noise and EMI protection. This configuration also illustrates that a star ground scheme isn't universally adopted and can be modified. As you can see, three ground connections are soldered to the back of the volume potentiometer, but the ground connection for the audio output jack is made at the tone potentiometer.

At first glance, the use of a potentiometer case for solder connections seems like poor construction technique. After all, a solder lug attached to the control plate would provide a neater, more professional layout. However, this is simply how it's done with legacy guitars from Fender, Gibson, and others. To guarantee authentic, vintage tone, the layout and components used in a stock Tele have remained essentially unchanged for decades.

Components

I don't want to give you the false impression that innovation in the design of electric guitars somehow halted in the 1950s or 1960s. To the contrary, there are numerous examples of electric guitars that push the envelope in electronics and materials, from the lightweight resin-based guitars from Parker (www.parkerguitars.com) and the MIDI-compatible guitars from Breedlove (http://breedlovemusic.com), to the synthesizer guitars from Moog (www.moogmusic.com) and Line 6 (www.line6.com). However, vintage-style guitars remain a significant force in the industry. Fortunately, if you learn the fundamentals of a vintage-style guitar like the Tele, you'll be well positioned to understand the more complex electronic guitar designs.

EMI Shield

Well-built electric guitars are designed to operate in electrically hostile environments—near other instruments, amplifiers, speakers, cables carrying power and audio, fluorescent lights with faulty ballasts, and cell phones. Of these signals, the 60Hz power grid is the most powerful, ubiquitous, and problematic. Without shielding, the wire loops within the magnetic pickups, the wires from the pickups to the controls, and wires to the audio output jack all form excellent antennas for receiving 60Hz hum.

As noted in the teardown, the designers of the Tele attempted to minimize hum pickup by painting the cavities with conductive paint and by grounding the large metal bridge plate. Painting guitar cavities with conductive paint is a common method of reducing noise, hum, and radio interference because it's effective and easy to apply. I use conductive paint from Stewart-MacDonald Instrument Repair Supply (www.stewmac.com), which contains graphite and carbon black. At $60 per pint, it's expensive but easy to work with. If you use it, don't touch the toxic paint while it's wet, and wash your hands after painting.

An alternative to painting is to line each cavity with self-adhesive copper tape. The adhesive backing is conductive, making overlapping seams electrically continuous. Because of the significant expense and time involved in working with copper tape, I prefer paint for cavities and conductive copper tape for the inner surface of the pickguard. Copper tape does look better than black paint in the cavities—a point to consider if you plan to show off your handiwork with a clear pickguard.

Wiring

In addition to using shielding to reduce hum, twisted-pair and coaxial cable can be used to connect the pickups, switches, controls, and output jack; the use of proper grounding techniques can also help. Twisted pairs of wires carrying signals from the pickups can minimize hum pickup, because the 60Hz signals induced in successive twists cancel each other. Coaxial cable minimizes hum pickup by enveloping the inner conductor with a grounded mesh or foil tube.

During the teardown, we saw that the ground wires from each cavity, the audio output jack, and the switch all terminated on the case of the volume pot. The use of a common point in the guitar to which all ground connections are tied is one way to minimize the chances that currents in one ground line will generate a signal in another part of the circuit sharing the same ground. In other words, if there are multiple grounding points, each ground is at a different potential, because there is a finite resistance from one ground point to the next. As such, if one ground path carries a significant current, a voltage drop occurs across the path, and this voltage can interfere with the signals from the magnetic pickups. It's important to note that using a common ground point isn't the only, or necessarily the best, way to insure proper grounding. However, it's tradition, and it usually works.

Another practice that is influenced by tradition is the use of fabric insulation. Some guitar builders insist on using 22AWG tinned, stranded wire insulated with a Celanese (acetate) wrap and waxed, braided cotton outside insulation instead of modern PVC. However, I'm not convinced that any human can hear a difference in tone because of the use of stranded versus solid core wire or of different types of insulation. PVC is less expensive, more stable in environments with varying temperature and humidity, and readily available.

Potentiometers

The Tele controls are 2W, 250KΩ audio taper potentiometers. The audio or log taper potentiometers approximate the logarithmic response of human hearing, so that a given rotation of the volume potentiometer results in equal changes in volume, regardless of the initial potentiometer setting. As discussed in previous teardowns, a linear potentiometer would provide a nonlinear change in volume that is dependent on the initial potentiometer setting.

A 2W potentiometer seems like overkill for a guitar that generates a few milliwatts. However, for many builders, tradition and vintage tone dictate the use of dated hardware. Also, from a practical perspective, the heat absorbed during the process of soldering multiple leads to the case of a metal potentiometer would probably destroy lesser components.

Capacitor

The stock tone control in my Tele uses a 250KΩ audio taper potentiometer and a 0.05μf, 100V ceramic disc capacitor. Ceramic disc capacitors are a good choice for this application because they're inexpensive, they're stable at a range of temperatures, and they fit nicely in the limited space of the control cavity. Mods commonly use different values of ceramic disc capacitors, sometimes in conjunction with one or more film capacitors.

Audio Jack

The 1/4-inch mono audio output jack is standard throughout the guitar industry. The jack is tough enough to withstand abuse from accidentally yanked chords and plugs, is relatively inexpensive, and requires little maintenance.

There are exceptions to the standard output jack. For example, Parker (www.parker.com), Paul Reed Smith (www.paulreedsmith.com), and a few other guitar manufacturers install stereo audio output jacks on their guitars that have both magnetic and piezo pickups. One channel is from the magnetic pickup and the other is from the piezo pickup and built-in preamplifier. In addition, some traveler guitars with built-in audio amplifiers sport 3.5mm jacks for use with lightweight headphones.

Switches

The traditional three- or four-position, spring-action lever switch, shown in Figure 12-6a, is used in virtually all Teles. Some non–Fender Telecaster models feature a modern, sealed lever switch. However, tradition and the pursuit of vintage tone often dictate the traditional, expensive, and somewhat bulky lever switch.

Alternatives to the lever switch include the open three-way toggle switch, shown in Figure 12-6b. This traditional switch, commonly used in Gibson guitars, while more compact than the lever switch used on the Fender Tele, is much larger than a modern, sealed toggle switch. A key advantage of an open toggle switch over a closed switch is that the contacts can be cleaned. Of course, a sealed switch is less likely to require cleaning.

Passive Magnetic Pickups

A guitar's tone is a function of hundreds of variables, from the thickness of the pick and composition of the neck, to the type of paint on the body. However, the magnetic pickups are the heart of the Tele's distinctive sound. They are also the most varied, most valuable, and least understood components on the Tele.

a

b

FIGURE 12-6 Four-way lever switch and three-way toggle switch

Passive magnetic pickups are available in hundreds of different models from dozens of manufacturers and an equal number of small artisan shops that cater to the custom market. Most of these pickups share the basic structure of one or more permanent magnets surrounded by a coil of copper. When a steel guitar string enters or vibrates within the magnetic field surrounding the magnet(s), a current is induced in the coil. The signal is routed to the controls, out the mono audio jack, and to the input of the audio amplifier.

Magnetic guitar pickups are based on the same effect used in dynamic microphones, in which a diaphragm moves either a magnet or a coil when sound hits the diaphragm. However, in the case of a guitar pickup, the physical plucking of a string provides the relative motion within the magnetic field.

From an electronics perspective, pickups are commonly classified as active or passive, single coil or double coil (humbucking), high or low impedance, high or low output, and magnetic or piezo. Pickups are further classified by type and configuration of the material(s) used in their construction, whether they are embedded in wax or epoxy, and whether the assembly is open or covered with a grounded shield. Now let's examine a passive magnetic pickup in detail with a mini-teardown.

Magnetic Pickup Mini-Teardown

You can follow the pickup teardown shown in Figure 12-7 at home if you don't mind destroying the pickup. A good pickup can account for up to a third or half of the cost of a guitar, and, other than requiring remagnetization every decade or two, they last forever. I frequently replace a pickup with one of a different design to change the tone of a guitar, but I carefully store the original pickup for future use.

Let's focus on the Tele's smaller, single-coil neck pickup. On inspection, you should see the six round ends of cylindrical pole pieces—one per guitar string. On a Fender pickup, the pole pieces are magnets. Other pickup designs use steel pole pieces to focus the magnetic fields generated by permanent magnets attached to the pole pieces. In either case, the movement of a string above a pole piece disrupts the large number of magnetic lines of force focused through each pole piece.

Let's examine the magnetic field surrounding the pickup. Because we can't sense the field directly, we have to rely on secondary indicators. The simplest approach is to use magnetic field viewing paper, which is a plastic sheet containing iron filings suspended in oil, available from Edmund Scientifics (www.scientificsonline.com). You should see a concentration of magnetic lines of force—that is, greater accumulations of iron filings—immediately above each pole piece.

An alternative to the above qualitative approach is to measure the magnetic field strength with a probe containing a Hall effect chip. The Hall effect is the production of a voltage difference across a conductor, transverse to an electric current in the conductor and a magnetic field perpendicular to the current. If you've used one of those clamp-on current probes, you've worked with a Hall effect probe designed to

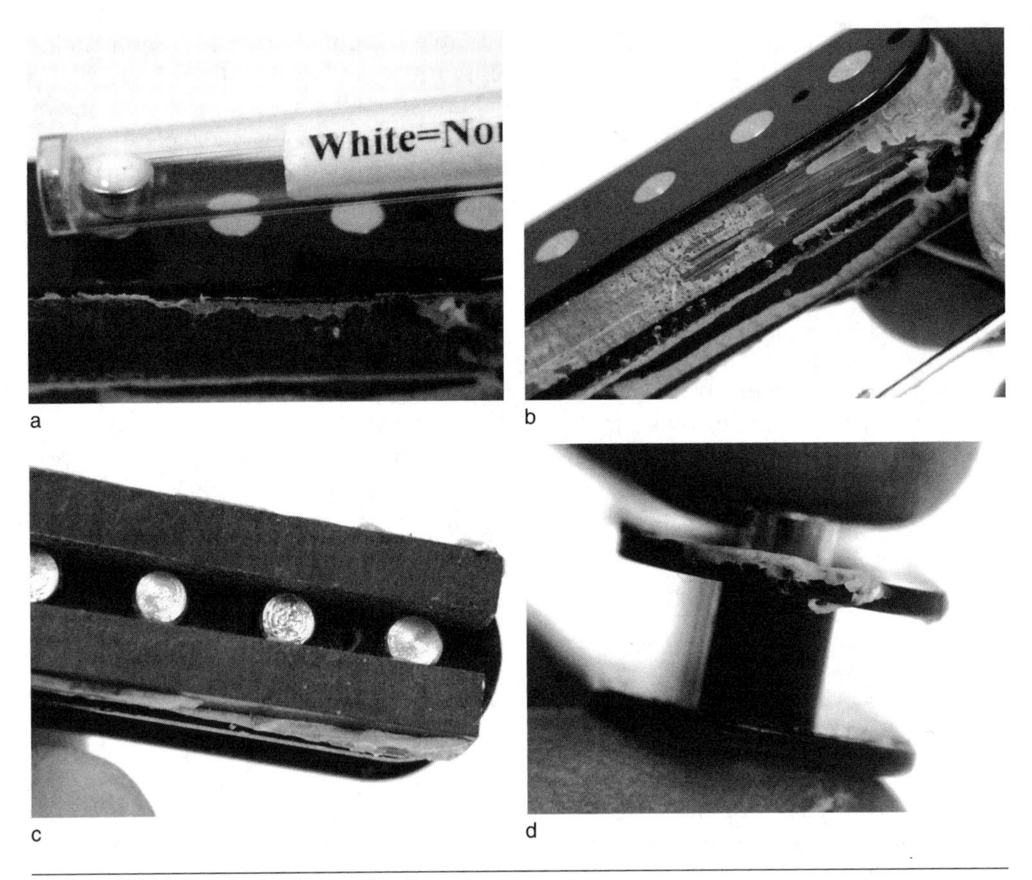

FIGURE 12-7 Mini-teardown sequence

measure current. Unfortunately, a precision Hall effect magnetic field probe, such as the B&K PR 26M (www.bkprecision.com), sells for thousands of dollars.

Regardless of how you measure it, what matters is that the stronger the magnetic field, the greater the voltage output from the pickup. Measuring the absolute field strength is more important with older pickups, because magnets degrade with time.

Step by Step

Because of the apparent simplicity of construction, you might get the impression that you're disassembling a simple solenoid or speaker coil. However, subtle differences in pickup construction can result in significant tone differences. For example, as you step through this mini-teardown, notice the windings. Are they regular within a layer? What about between layers?

Step 1

For our purposes, the most relevant magnetic field parameter is polarity, and this is relatively easy to measure. I use a commercially available detector from Stewart-MacDonald—a magnetized steel ball with one pole painted black and one white, contained within a plastic tube. A small keychain compass works almost as well. With your magnetic polarity detector, determine the polarity of each of the pole piece slugs, as shown in Figure 12-7a. If you have a modern Fender Tele, both the bridge and neck pickups should be "south up," where "up" means facing out, toward the steel guitar strings. If you're lucky enough to have a Fender Tele from the early 1950s, the neck pickup should be "north up."

The relative magnetic polarity of each pickup determines the phasing of the pickups, and therefore how the white and black output wires should be connected to the controls for a proper, in-phase signal. If you wire your neck and bridge pickups so that they're out of phase, the signals will partially cancel each other. Each pickup will sound fine by itself, but when the selector switch connects the pickup wires in series or parallel, the resultant output signal will be weak and tinny sounding.

Documenting a pickup's magnetic polarity is useful beyond verifying proper phasing. Polarity also indicates whether the original pickups have been modified—an important consideration when you're verifying the authenticity of a guitar at a garage sale or a guitar shop. For example, if you have been following along closely, you can see in Figure 12-7a that the polarity of the pickup is south up. This is inconsistent for a modern Fender pickup.

There's a good reason for this discrepancy. The pickup in the photo is a single-coil neck pickup from a Gibson Melody Maker that is physically similar to a Fender Tele pickup, but it uses a reverse polarity. I couldn't bring myself to destroy a perfectly good (and expensive) Fender pickup, when I had this relatively inexpensive Gibson pickup to use for a teardown. Besides, even if I had used the stock Fender pickup, it may not have resembled the neck pickup of your Tele if you're following at home.

Pickups used in the Tele differ by model and year of manufacturer. For example, some Tele neck pickups are encased in a metal shield that is connected to ground. Others' designs shield the pickup from 60Hz hum by wrapping the pickup with insulating cloth tape and then wrapping the coil with conductive copper tape, which is then connected to ground. In our teardown, we have a simple, single-coil pickup without a metal shield or copper tape.

Step 2

Now, remove the black paper or cloth tape from the pickup, exposing the 7800 to 8000 windings of fine enameled copper wire, as shown in Figure 12-7b. Tele neck pickups are typically constructed of fragile 43AWG enameled wire, whereas 42AWG wire is used for the bridge pickup. The 43AWG wire is used in the neck pickup because it is shorter than the bridge pickup—the guitar strings are closer together near the neck and wider near the bridge. To get enough windings around the neck pickup, the manufacturer uses thinner wire. The white substance on the surface of the windings in the photo is wax, which minimizes vibration and seals the unit from moisture.

If you have a stock Fender pickup, such as the Vintage Noiseless pickup, the six pole pieces are Alnico (aluminum, nickel, cobalt) magnets. However, if you're following along at home with a non-stock Tele or another guitar, you might encounter one of dozens of possible pickup designs. For example, your pickup may include two rectangular Alnico bar magnets that run on either side of the pickup, as in Figure 12-7c. The magnets make contact with each of the six steel pole pieces. Remove the bar magnets, if present.

Step 3

Remove the wire from the pickup to reveal the plastic bobbin, as shown in Figure 12-7d. If you want to save the wire by unwrapping it, give yourself an hour or two to complete the task. And you're likely to break the fragile 43AWG wire. Alternatively, attack the wire with sharp diagonal cutters and dispose of the short wire clippings. If you're interested in winding your own pickups, 42AWG and 43AWG enameled wire is readily available from several guitar supply houses.

Although the wire used to create magnetic pickups is fragile, it enables pickup designers to pack a lot of inductance into a small space. Consider the Fender Vintage Noiseless pickups, which are single-coil pickups based on Alnico 5 magnets. The 3.7H neck pickup and larger 4H bridge pickup contribute to the Tele's characteristic twang. Unfortunately, the single coil pickup design also works wonderfully as a 60Hz antenna.

Pickup manufacturers address the hum problem by using humbucker pickups, which are designed with two coils of wire and two magnets. A humbucker is constructed so the two coils of wire are out of polarity with each other and each coil is wound around a magnet of opposite polarity. Because the coils are in close proximity to each other, 60Hz currents induced in one coil are induced in the other coil. However, because the coils are wound in opposite polarity, the 60Hz signals are 180 degrees out of phase, canceling the hum.

Consider the contribution of the magnets of opposite polarity. The 60Hz currents are induced by an external electromagnetic field and are not affected by the presence of the magnets. However, because the magnet polarities are reversed, the voltage induced by a vibrating guitar string is doubled. Humbucking pickups not only increase the SNR of a guitar's pickup system, but they create a thicker tone than single-coil pickups. As such, if you replace the Tele's single-coil pickup with a humbucker—a common mod—you'll lose some of the Tele's characteristic biting tone. And that may be a good thing. As with most mods that affect a guitar's tone, it's a question of personal taste.

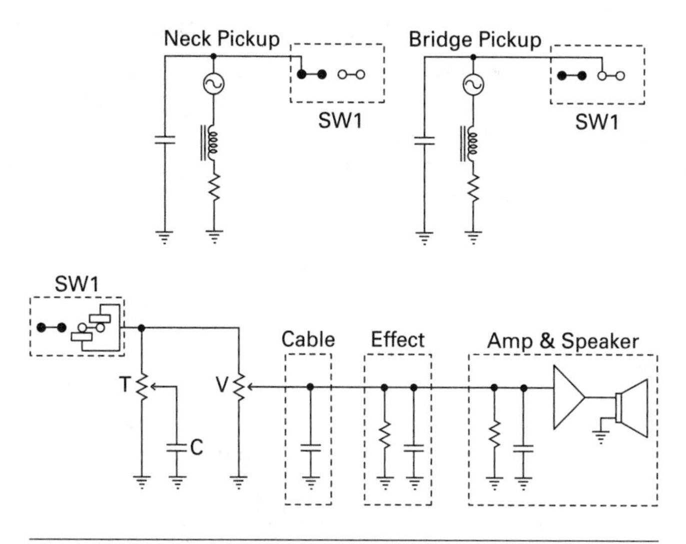

FIGURE 12-8 Simplified schematic of the Telecaster and typical setup

How It Works

An electric guitar is like a CD or DVD player in a home entertainment system, in that the listening experience depends as much on the speakers, amplifiers, mixers, and other devices as it does on the signal source. As such, in considering the operation of our Tele, we have to assume a typical system configuration—at least one cable, perhaps an effects pedal, and an amplifier with a built-in or external speaker system, as in the schematic shown in Figure 12-8. For our purposes, the Tele can be reduced to a magnetic neck pickup, magnetic bridge pickup, pickup switch (SW1), tone potentiometer (T), volume potentiometer (V), and tone capacitor (C).

The electrical equivalent of a pickup is an AC signal source, series resistance, inductance, and parallel capacitance. Recall from our teardown that no resistors or capacitors are associated with a passive magnetic pickup. The resistance is due to the DC resistance of the fine copper wire—typically around 10KΩ for a single coil Tele pickup. The capacitance is the equivalent capacitance between coils of wire. Most references that I've seen suggest that a normal range for coil winding capacitance is 100 to 200pF. As noted earlier, the published inductance values for single coil pickups range from 2 to 5H.

The significance of the equivalent circuit for each pickup is that they appear as classic parallel LC circuits. Recall that at resonance, a parallel LC circuit presents a high impedance and corresponding high voltage across the inductor and capacitor. The lower the value of internal resistance, the sharper the peak of resonance. Conversely, with increased resistance, the resonance peak becomes less pronounced.

If we assume a Fender Vintage Noiseless neck pickup with an inductance of 4H and an equivalent capacitance of 160pF, then the resonant frequency of the neck pickup in free space can be calculated as follows:

$$F = 1 / [2p\ (L\ C)1/2]$$

$$F = 1 / [6.28 \times (4 \times 160 \times 10^{-12})^{1/2}]$$

$$F = 6291\text{Hz}$$

Similarly, the theoretical free space resonant frequency of a 3.7H bridge pickup with an equivalent capacitance of 160pF can be calculated as follows:

$$F = 1 / [6.28 \times (3.7 \times 160 \times 10^{-12})^{1/2}]$$

$$F = 6541\text{Hz}$$

If you don't have a calculator handy, a nice LC resonance frequency calculator is available at What Circuits (www.whatcircuits.com). Try plugging in values for the range of equivalent capacitance from, say, 50 to 200pF, and notice the resonant frequency.

Returning to the schematic in Figure 12-8, you'll see that the pickups are not in free space. They can be connected in parallel with each other and with the other systems' components. The capacitance across the inductance of each magnetic pickup includes a contribution from the 0.05µf tone capacitor (C) as well as the equivalent capacitance from the cable, effects pedal, and amplifier input circuitry. The typical cable contributes 200 to 500pF, depending on length and manufacturer. Cable resistance is insignificant, given the low signal levels involved. Effects pedals easily contribute another 100 to 200pF to the system. In addition, pedals that don't completely disconnect when powered off contribute as little as 500KΩ of shunt resistance. The typical amp input also contributes about 1M of shunt resistance and 100pF capacitance. All this has an effect on the tone of the system, namely shunting high frequencies to ground.

The resonant frequencies of the pickups are significant in that they're within the most sensitive band of human hearing. As such, the nonlinearities in frequency response they introduce into the electric guitar system will likely be evident to the listener. However, it's difficult to quantify how the other system elements connected to the guitar can shift the resonant frequency of the system and therefore the tone. You could use SPICE (Simulation Program with Integrated Circuit Emphasis) or a similar tool to simulate the circuit, but you'd be hard pressed to provide the exact electrical parameters of your specific setup—your guitar and pickups, your cables, your effects pedals, and your amp.

For example, it's almost impossible to find capacitance and resistance loading values for effects pedals that don't totally disconnect from the circuit when not in use. And cable manufacturers like to use qualitative superlatives in defining their guitar cables, but few list actual specifications. One exception is Lava Cable (www.lavacable.com), which manufactures cables with capacitance of 25 to 50pF per foot.

To get a feel for the relative contribution of components, let's run a few experiments with the aid of a signal generator and oscilloscope. Since we know the current through a parallel LC circuit will be minimum at resonance, we can apply a constant amplitude audio signal across a pickup and sweep the frequency. A dip in current should indicate resonance.

As shown in the plot in Figure 12-9, the resonance dip in current for the Fender Vintage Noiseless neck pickup is about 3kHz, which is significantly different from the free space resonant frequency calculations. Differences in theoretical and measured values could be due to differences between the actual versus published inductance values and assumed capacitance values. Notice that for illustration purposes, I cleaned up the actual plot, which wasn't as pronounced or clean as that shown in the figure. The change in current at resonance is only on the order of a few milliamps with a 5V sweep generator.

To determine the resonant frequencies of the guitar as a whole, you could use the same method of sweeping a constant amplitude audio signal, but this time with the pickups mounted in the guitar body, with the tone capacitor in the circuit. The audio signal could be input and measured at the audio jack. Similarly, you could measure the resonant frequency of your system as a whole—say, for example, the Tele body, a 30-foot Fender coil cable, an MTX distortion pedal, and a Fender Champion tube amp. You could then substitute components and measure the resonant frequency.

At issue, of course, is the value of these measurements beyond our interest in electronics. We can't assume that our objective measures will reflect the quality of tone produced by the system, and therefore the value to the guitar designer or player. Perhaps you can now appreciate why Fender hasn't changed much in its original Tele design, including the use of antiquated components and wiring techniques. This isn't to say that the guitar shouldn't be modded for fear of

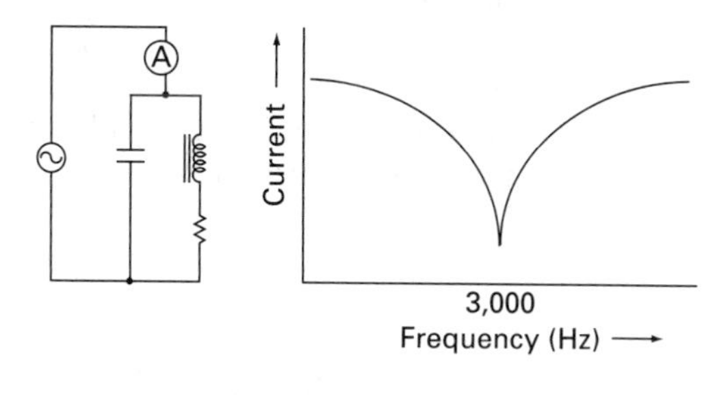

FIGURE 12-9 Idealized dip in current at resonance

introducing some unknown variable into the system and wrecking the tone. On the contrary, the beauty of the Tele is that the design is so simple that most mods can be undone in a few minutes, without any permanent damage.

Mods

Despite the Tele's apparent mechanical simplicity, you might be amazed at the tonal differences that can result from subtle changes in components and wiring. In determining which of the following mods are for you, it's worth repeating that tone is a matter of taste and music style. A stock Tele may be all you ever need—but then perhaps you won't be able to resist manipulating the electronics to create a better tone. If so, here's a list of popular electronic mods for the Tele.

More Pickup Switching Options

One of the most popular and inexpensive mods is adding a serial pickup option by swapping a four-way selector switch for the stock three-way switch. The serial, in-phase tone is distinctive and useful with a variety of styles.

Compensated Volume Control

At low volume settings, the highs in the guitar signal are attenuated disproportionally. A work-around is to set the volume potentiometer to maximum volume and adjust overall volume by adjusting other system components. A popular mod is to use a treble bleeder capacitor—a 0.001µf ceramic disc capacitor—from the high side to the wiper arm of the volume pot. The capacitor provides a low impedance path for high frequencies, regardless of the pot's position.

Functional Tone Control

As noted earlier, a shortcoming of the Tele's design is the suboptimal tone pot. It's a common practice to set the potentiometer to 250KΩ for maximum treble and to forget about it. However, even at maximum, high frequencies are attenuated. For example, at 3000Hz, the total impedance to ground is essentially the value of the pot, given the capacitive reactance is negligible:

$$\begin{aligned} \text{Impedance} &= R + X_C \\ &= 250K\Omega + 1/[2\pi \times F\ C] \\ &= 250K\Omega + 1/[6.28 \times 3000 \times 5 \times 10^{-8}] \\ &= 250K\Omega + 1.1K \end{aligned}$$

The obvious way to maximize the high frequency output of the Tele is to use a higher value tone pot. Substituting a 500KΩ or even a 1MΩ potentiometer for the 250KΩ potentiometer is a common mod.

A commercial mod that does away with the tone potentiometer and 0.05µf tone capacitor, Stellartone, is based on switched capacitors instead of a pot. I've used the Stellartone (www.stellartone.com) on two Teles with good results. However, as with the standard tone pot, I tend to use it wide open, and I use an effects pedal and amp controls to vary the relative contribution of highs to the final tone.

Pickup Variations

You probably don't have enough time or money to try all the various pickups available for the Tele. In this teardown, we touched on the most popular configurations of passive magnetic pickups. In addition to a variety of passive magnetic pickups, there are active magnetic pickups and piezo pickups, both of which may have onboard amplifiers and impedance-matching circuitry that require battery power. Pickup variation mods range from using a different variety of passive pickups to a combination of active magnetic and piezo pickups with a passive magnetic pickup.

Passive Magnetic Pickups

In addition to the stock pickups from the big-name guitar manufacturers, several third-party manufacturers specialize in passive magnetic pickups for specific guitars. Figure 12-10 shows a popular replacement neck pickup for the Tele designed by Kent Armstrong (www.kentarmstrong.com). The pickup, produced in Korea, is more expensive than the equivalent Fender pickup, but it's more affordable than the hand-built pickups available from boutique shops. If you look carefully at the figure, you can see four leads—one pair for each coil in this hum-canceling pickup.

FIGURE 12-10 Kent Armstrong pickup

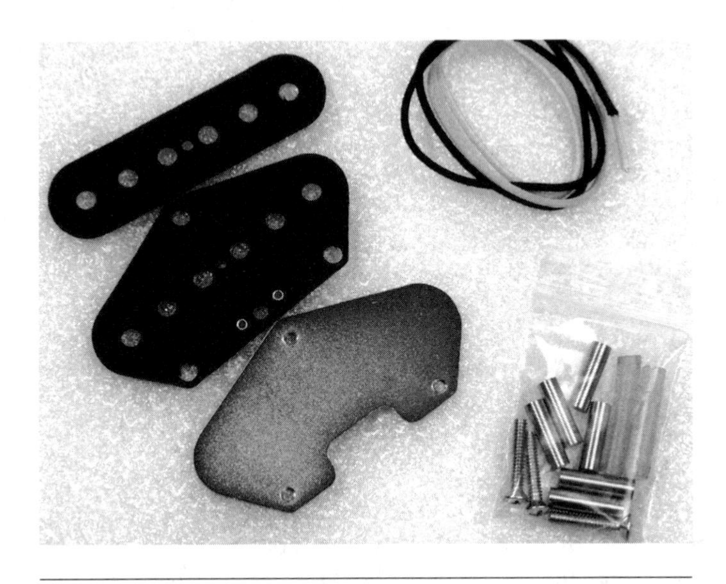

FIGURE 12-11 Single-coil kit

If you can't find or afford the passive magnetic pickup you need, or you simply can't resist working with 43AWG wire, then another option is to wind your own. Figure 12-11 shows a single-coil kit for the Tele that I purchased from Steward-MacDonald (www.stewmac.com).

The kit comes with Alnico 5 magnet pole pieces and the plates to hold the pole pieces in place. You'll need some 43AWG wire and steady hands to complete the pickup without kinking or snapping the wire. You might as well invest in a full roll of wire at the start, given your only option in dealing with a broken wire is to start over. Even if you don't build your own passive magnetic pickup, it's worth visiting the Stewart-MacDonald site to review the specs for common pickups, including the number of turns of wire, magnet polarity, and type of magnet.

Active Magnetic Pickups

Passive magnetic pickups are limited in the signal strength they can achieve. Increase the strength of the magnets and the steel strings are pulled into the pickup, damping the notes. Move the strings closer to the pickup and you have the same problem. One way to address this and other shortcomings of the passive magnetic pickup is to install an amplifier and interfacing circuitry at the point the signal wires exit the magnetic pickup coil. The leader in the active magnetic pickup market is EMG (www.emginc.com).

Active magnetic pickups are considered a "hot" signal source, with a typical output of 1.5VAC. In addition, they are usually shielded internally. Both factors contribute to an SNR that is superior to that of a typical passive magnetic pickup.

Because active pickups have little magnetism compared to passive magnetic pickups, strings can be positioned much closer to the pickup, which can increase playability.

Aside from cost—about double that of a passive magnetic pickup—the major downside of using an active magnetic pickup is the need to supply power, typically 9VDC at 80μA. The issue is typically lack of space for a 9V battery. My suggestion is to use a router to create a cavity in the guitar back and install a battery holder. Steward-MacDonald and WD Music Products (www.wdmusic.com) sell battery holders and router guides. In addition, if you upgrade to active magnetic pickups, you'll have to replace the standard 250KΩ potentiometers with 25KΩ potentiometers.

Piezo Pickups

Piezoelectric, or piezo, pickups are vibration sensors. They're standard issue on most acoustic-electric guitars and increasingly common on hybrid magnetic-piezo electric guitars. Piezo pickups rely on the piezoelectric effect, in which a voltage is produced when a crystal is stressed. Because of their structure, piezo pickups are relatively impervious to magnetic fields, including those that cause hum. The high impedance output of a piezo pickup can be sent directly to an output jack. However, it can't be mixed with the output from a magnetic pickup without an active mixer/preamp.

Properly configured, a piezo pickup can transform a Tele or another electric guitar into a semi-acoustic guitar with shimmering highs. Moreover, with an onboard mixer, the high impedance signal from a piezo pickup can be combined in varying degrees with the signal from an active or passive magnetic pickup, resulting in unique tone combinations somewhere between an acoustic and a traditional Tele.

Examples of piezo pickups compatible with the Tele include the GHOST pickup system from Graph Tech (www.graphtech.com) and the Tune-O-Matic Powerbridge from Fishman (www.fishman.com). Onboard piezo mixers and amplifiers are available from L.R. Baggs (www.lrbaggs.com) and Fishman, among others. These amplifiers and mixers provide isolation and impedance matching for the high-impedance piezo pickup, and typically provide auto-mixing as dictated by a smart 1/4-inch output jack. If the jack senses a stereo plug, magnetic signals are routed to one channel and the piezo signals to another. If it senses a mono plug, the mixer combines the signals from the piezo and magnetic pickups, with the relative contribution of each signal controlled by the position of the mixing pot.

A MIDI Interface

Transforming your Tele into a MIDI controller opens up myriad possibilities, from automatic transcription of music notation to controlling drum machines and synthesizers. You can add a MIDI interface to your Tele in two ways—the easy way and the hard way—and I've done both. The easy way, and the way I recommend if you're new to MIDI, is to purchase an external MIDI pickup, such as the Roland GK-3 (www.roland.com), shown in Figure 12-12.

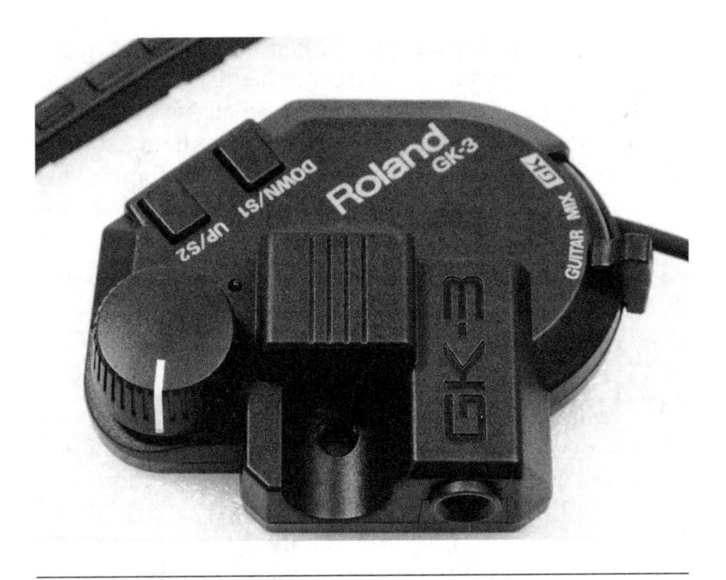

FIGURE 12-12 Roland GK-3 MIDI pickup

FIGURE 12-13 Pickup installed in reworked pickguard

To install the unit, mount the thin, active magnetic pickup immediately above the bridge plate and attach the GK-3 unit with either temporary adhesive or two wood screws, as shown in Figure 12-13. Notice that the pickup is composed of six separate pickups, as opposed to a single pickup with six pole pieces. Each of the six pickups responds only to the string immediately above it. In this way, the signals are easily parsed and it's possible to determine which string contributes a specific note during play.

Ideally, the active magnetic pickup should be mounted as close to the bridge as possible. The design of Tele's bridge plate is such that you have to mount the pickup just above the plate. However, at this position, the pickguard presses the pickup against the strings. The solution is to cut the pickguard with a jigsaw, coping saw, or band saw to make room for the pickup. Consider purchasing an inexpensive pickguard to cut and use with the MIDI pickup. You can reinstall the original pickguard if you decide to remove the pickup.

The hard—and most attractive—way to attach a MIDI interface to a Tele is to use the embedded version of the Roland pickup. Instead of working with double-sided tape, however, you'll need a drill, soldering iron, and the better part of a day. You'll need to drill three holes in the pickguard and in the body for switches. In addition, you'll have to add a third potentiometer between the tone and volume controls. More significantly, you'll have to bore a 1-inch diameter, 2-inch deep hole from the edge of the guitar to the control area for the 13-pin MIDI connector. And, to add an extra challenge to the effort, the three ribbon cables included in the standard kit aren't long enough for the Tele. You'll either have to extend the cables or find longer versions at an electronics supply house.

I didn't want to wait a week for delivery of longer cables, so I opted for extending the original cables by splicing in a few extra inches of ribbon cable. Fortunately, the operation was successful. With the exception of the thin magnetic pickup mounted just above the bridge, my Tele looks and handles like a standard electric when I'm not using it as a MIDI controller. Next time, however, I'll buy cable extensions and save a few hours of bench time.

Chapter 13

Effects Pedal

Guitarists who strive for a particular tone for each song or genre they play have two options: they can buy a Fender Stratocaster and a Marshall amp to play one song, a Gibson Les Paul and VOX amp to play another, and so on; or they can fake it with signal processing. Few musicians have the funds or space for the first option, making analog and/or digital signal processing the de facto means of morphing sounds.

Signal processors for guitars are available in a variety of configurations. They can take the form of a chip embedded in a guitar body or amplifier, a rack-mounted unit, or an effects pedal. An effects pedal, such as the one shown in Figure 13-1, is a signal processor in an indestructible case with a rugged on–off button. Effects pedals

FIGURE 13-1 MXR Distortion+ effects pedal

are designed to interface with the musician's shoes and feet. The on–off buttons are stomped, and guitarists operate rubber-encased knobs with their feet—all without interrupting guitar play.

Effects pedals are available for almost every conceivable sound-processing task, from creating a classic tube sound to adding reverb, chorus, wah, or compression to the signal from a guitar. With a multi-effects pedal or handful of analog or digital single-effects pedals, a skilled guitarist can play authentic-sounding acoustic, hard rock, metal, country, jazz, and blues with a single electric guitar and amp.

In this chapter, I'll tear down a popular analog effects pedal, the Dunlap M-104 MXR Distortion+ pedal, referred to hereafter as simply "the pedal." As its name suggests, this popular signal processor introduces a controllable level of distortion into the guitar signal. The simplicity of the pedal's circuitry lends itself to modding, which I'll cover at the end of this chapter.

Highlights

Give yourself 10 minutes to complete the basic teardown. The pedal consists of an aluminum box that contains a well-laid-out circuit board with about two-dozen components. There's a 741 operational amplifier, a pair of 1/4-inch audio jacks, two audio-taper pots, switching diodes, metal film resistors, an LED, a variety of capacitors, and a foot switch.

During the teardown, note the following:

- **External power requirements** Effects pedals, like most other music electronics, use positive ground instead of conventional negative ground.
- **Layout of the circuit board and overall construction** Effects pedals are built to take abuse.
- **Component types and ratings** Try to date the design of the board, based on components.

Specifications

Effects pedal specifications are often written in qualitative music jargon. For example, Dunlap's materials for this pedal refer to "retro-authentic soft-clipped distortion tones" and "classic fuzz tones." While these terms may be understood by a guitarist, they lack the technical specificity required for an objective comparison of pedals from different vendors.

Many specifications of industry standards are simply understood. For example, virtually all effects pedals are configured so that the audio input jack is on the right side; the audio output jack is on the left; and the controls knobs, LED, and on–off button or pedal are on the top. Figure 13-2 illustrates this standard pedal configuration.

The effects pedals on either side of the Distortion+ pedal in Figure 13-2 are two of my favorites from Roland/Boss, a major manufacturer of single- and

FIGURE 13-2 Standard pedal configurations

multiple-effects pedals. The Roland/Boss OD-2 Turbo Over Drive pedal is similar to the Distortion+ in how it modifies the guitar's signal, but it uses three times as many discrete components. The BD-2 Blues Driver pedal creates a much richer distortion than either of the other pedals, but it uses a high discrete component count.

As you can see from Figure 13-2, an advantage of a common right-input, left-output design is that pedals can be cascaded with short, direct cables. Typical effects pedals, including the three in the figure, present a nominal input impedance of 1MΩ and a nominal output impedance of about 2KΩ, and they require 9VDC (positive ground) at up to 500mA.

Operation

Since this is a stand-alone effects pedal with only two controls and an on–off button, as shown in Figure 13-2, operating the pedal is a no-brainer. Insert a 1/4-inch mono audio plug from your guitar or effects pedal into the input jack, and connect the mono plug to another effects pedal or a power amp into the output jack. Stomp on the large button to activate the circuitry, including the LED power indicator. Turn the DISTORTION knob clockwise to increase the amount of distortion in the signal. You can turn the OUTPUT knob to alter the volume of the guitar signal, without affecting the level of distortion.

Pedals are typically used as part of a sequence of five or six pedals. A typical pedal setup might include, from guitar to amp, a wah-wah pedal, chorus pedal, overdrive pedal, and delay pedal. Because of the interaction among pedals, the same pedals can be arranged to provide a variety of tone.

For example, if you place a reverb pedal before a distortion pedal in the signal chain, you'll get a very different tone compared to using a distortion pedal and then a reverb pedal. And substituting a pedal with a specific effect from, for example, Roland/Boss with one from Ibanez will result in a different tone.

Because of expense and complexity of using multiple, single-function effects pedals, digital, all-in-one, multi-function pedal boards are used by some guitarists. However, dedicated, single-function effects pedals set the standard for guitar signal processing.

Teardown

The teardown is illustrated in Figure 13-3. As you can see, it's a 5- to 10-minute operation. Even if you're following along at home with an effects pedal other than the Distortion+, once you remove the back and expose the circuit board, you're 80 percent there. The remaining 20 percent of effort is in analyzing the circuitry on the board.

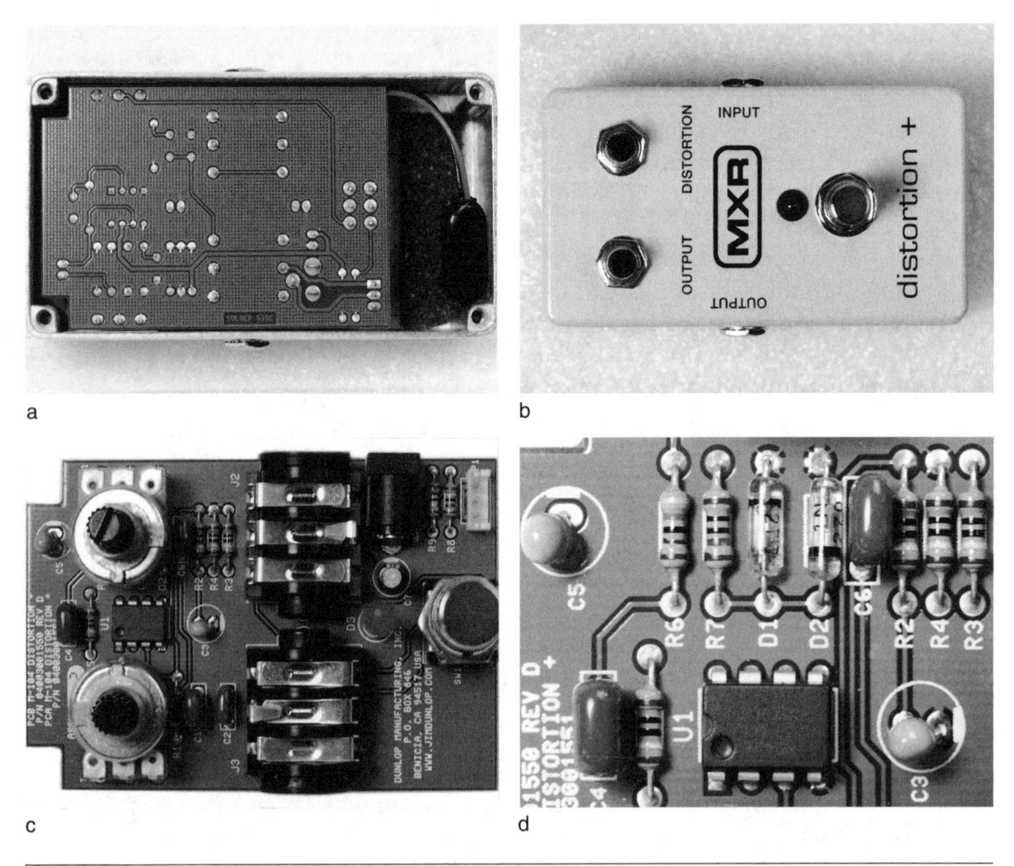

FIGURE 13-3 Teardown sequence

Hundreds of different effects pedals are on the market, and each manufacturer has its own way of achieving a particular tone. For example, the popular Roland/Boss distortion pedals use half a dozen transistors, whereas our pedal achieves similar tones with a single operational amplifier. The distortion pedals from different manufacturers produce similar but different tones. There is no clear-cut better tone—it's a mater of personal taste.

Tools and Instruments

A Phillips-head screwdriver, nut driver set, and crescent wrench are all you need for this teardown. If you intend to make a habit of removing retaining nuts from switches and buttons mounted on front panels, consider picking up a plastic nut removal tool from Steward-MacDonald (www.stewmac.com). Other pedals may have pressure screws in their plastic control knobs, requiring either a small Allen wrench or small flat-blade screwdriver.

A multimeter might help in tracing the schematic, but it's not necessary. A dual-trace oscilloscope would enable you to visualize the effect of the pedal at various distortion settings.

Step by Step

This teardown involves opening the metal box and freeing the controls. However, take your time. It's easy to scratch the paint on an effects pedal when you're removing the retaining nuts on switches and potentiometers.

Step 1

Remove the back cover. Extract the four machine screws with a Phillips-head screwdriver and set aside the back cover, shown in Figure 13-3a. If a 9V battery is installed, remove it from the battery cavity.

Step 2

Remove the external retaining nuts. Carefully remove the retaining nuts from the external jacks, switches, and potentiometers, which are visible in Figure 13-3b. You'll need a 7/16-inch nut driver for the input and output jacks, and a 13mm deep metric driver for the on–off button. Use the plastic nut removal tool to avoid scratching the paint when you remove the retaining nut from the on–off switch. The metal retaining nuts screw into the soft plastic of the input and output jacks—so avoid cross-threading the jacks when you reseat them.

Step 3

Extract the circuit board. Remove the 9V battery connector and lead from the board to avoid damaging the wires.

Step 4

Examine the board. The two potentiometers, input and output jacks, and push button switch dominate the circuit board, as shown in Figure 13-3c. Note the operational amplifier and associated components, shown in Figure 13-3d.

Layout

Thanks to the use of low-density leaded components, the layout of the circuit board is clean and the connections are easy to trace with the aid of a multimeter. As a bonus, component labels are stenciled on the board. Carefully bend the potentiometers out of the way and orient the board so that the on–off button is at the right, as shown in Figure 13-4. From this perspective, the components associated with the power circuitry—the power adapter jack, 100µf at 16V electrolytic capacitor, and 9V battery jack—are at the upper-right. The power-on LED is just to the left of the large button.

The normal signal path is from the right jack to the left jack, which is from top to bottom in this orientation. The jack on the top is a special sensing jack that cuts power to the unit when a plug is removed. The signal processing components, including a 741 operational amplifier, a pair of germanium diodes, 0.25W metal film resistors, and several capacitors, are in the upper-left area of the board. See Figure 13-5 for a close-up of the two germanium diodes and operational amplifier.

FIGURE 13-4 Component layout

FIGURE 13-5 Germanium diodes and operational amplifier

Components

The components in this effects pedal are hardly representative of the cutting edge of electronics. As with many guitars, you'll find the use of traditional, leaded components common in many effects pedals designed to provide a vintage tone.

Diodes

The centerpiece of the pedal is a pair of 1N270 germanium diodes, labeled *D1* and *D2* in Figure 13-5. Unlike typical silicon power diodes used in power supplies, germanium diodes are relatively fragile, and their uses are limited to low-power switching applications, RF input probes, and AM/FM detectors. The 1N270 can handle an average forward current of only 40mA and a peak inverse voltage of only 100V.

For our purposes, the most significant characteristic of germanium diodes is their relatively low forward conduction voltage. As shown in Figure 13-6, the relationship between forward voltage and current is nonlinear for both germanium and silicon power diodes. With increasing forward voltage, there is an approximately exponential increase in current flow. Figure 13-6 also illustrates that germanium diodes conduct when forward-biased about 0.3V, versus 0.7V for silicon diodes.

Diodes are "soft switches," in that they don't instantly transform from perfect insulator and perfect conductor at a specific forward conduction voltage. Figure 13-6 illustrates that both standard silicon and germanium diodes conduct in varying

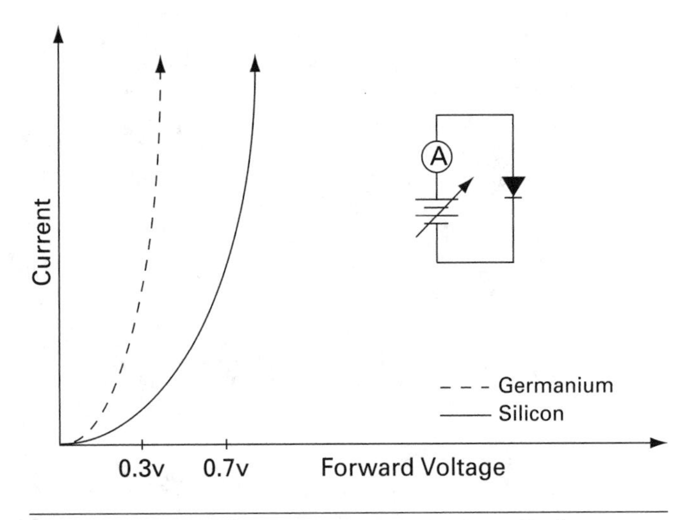

FIGURE 13-6 Silicon versus germanium diode conduction

degrees at 0.3V. Whereas the germanium diode is at nearly full conduction, the silicon diode is barely conducting. Both standard silicon and germanium diodes present high impedance to current when reverse-biased—until their peak inverse voltage is exceeded and an avalanche of current ensues.

The forward-bias conduction voltages of diodes are additive, in that if two germanium diodes are connected in series, cathode-to-anode, the series will conduct when forward-biased at 0.3V + 0.3V = 0.6V. Similarly, a germanium diode in series with a standard silicon diode will conduct when forward-biased at 0.3V + 0.7V = 1.0V.

You may be wondering why I keep referencing "standard" silicon diodes. I do this because a special class of silicon diodes called *Schottky* or *hot carrier* diodes are used for high-speed switching. The forward voltage drop is typically between 0.15 and 0.45V and the on–off switching time is often an order of magnitude less than that of standard silicon diodes. However, most effects pedals use either standard silicon or germanium diodes, so we'll focus on those technologies for now.

The fixed forward conduction voltage of a diode or series of diodes can be applied to a variety of applications. One use is to create a constant voltage drop, independent of current. For example, to run a 5.0V circuit from a 6.0V battery, you could insert a silicon power diode in series with the battery, providing the circuit with 6.0V – 0.7V = 5.3V. This may be within the maximum voltage limits for the circuit. Alternatively, you could connect two diodes in series, providing a working voltage of 6.0V – (0.7V × 2) = 4.6V, again, assuming this voltage will work for the circuit.

Another use for the forward-bias voltage drop is to create a clipper or limiter to constrain the amplitude of an AC signal. Within the power-handling capabilities of a diode, when a voltage in excess of the forward conduction voltage is applied to a diode, the device appears as a low-resistance shunt. A series resistor can extend the operating range of a diode clipper or limiter by protecting the diode from excessive current.

A single germanium diode will limit the amplitude of an AC signal applied across it to 0.3V, but only during the part of the cycle when the anode is positive with respect to the cathode. Virtually no current flows when the diode is reverse-biased. If symmetrical clipping is required, two diodes can be used in parallel, arranged anode-to-cathode. When one diode is forward-biased, the other is reverse-biased.

Circuit Board and Enclosure

Effects pedals are known as "stomp boxes" for good reason—they're designed to take abuse. Unlike many other electronics devices, they're intended to be stepped on repeatedly, to withstand an occasional dousing with soda or beer, and to have their power and signal chords yanked out mercilessly. As a result, the circuit board, jacks, and all-metal enclosure of this pedal are substantial. The aluminum enclosure not only provides structural integrity and a hefty heat sink, but also EMI shielding to and from other electronic devices.

The circuit board is also multilayered, meaning a layer of conductive traces are sandwiched between the top and bottom layers. While a multilayered board makes tracing connections between components more challenging, it often includes a large grounded area that shields the circuit from noise.

Resistors

This pedal uses a 47KΩ audio taper pot for the output control, a 470KΩ audio taper pot for the distortion control, and about a dozen metal film fixed resistors. Metal film resistors are appropriate for effects pedals in general because they produce less noise than cheaper carbon resistors. Resistor noise introduced at this point in the signal chain may be amplified several-fold by other pedals and the power amp.

This pedal is a low-power device that could have been produced with SMT resistors and other components. However, given that the size of the enclosure is already as small as practical, and that the input and output jacks need a substantial board for physical support, there is no obvious reason to develop a more compact board.

Operational Amplifier

This pedal is built around the popular LM741CN, a contemporary version of the 741 operational amplifier, or op amp, introduced in the 1960s. This ubiquitous

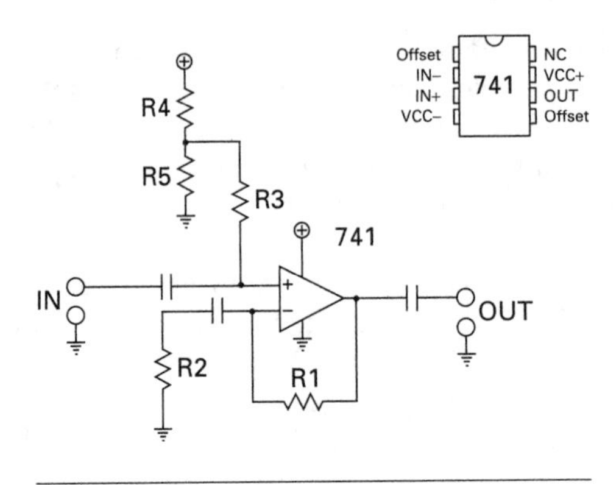

FIGURE 13-7 Noninverting AC operational amplifier circuit

component, containing 20 transistors and about a dozen resistors, has several features that make it ideal for use in an effects pedal. It can be configured as a high-input impedance, low-output impedance amplifier, and gain can be easily adjusted using a few resistors in a feedback network. Moreover, the bandwidth is more than adequate for audio work.

Figure 13-7 show a generic noninverting AC amplifier based on the 741 operational amplifier. The operational amplifier has two inputs, a noninverting input (+) and an inverting input (–). Because the guitar input signal is fed to the noninverting input, the output will be in phase with the input. Conversely, if the input signal is sent to the inverting input (–), the output will be 180 degrees out of phase with the input. Both inputs are critical to manipulating the behavior of the operational amplifier through the external feedback provided by the voltage divider formed by R1 and R2.

My favorite heuristic for working with operational amplifiers, borrowed from *The Art of Electronics* by Horowitz and Hill, is this: "The output will tend to change in a way that forces the external feedback network to minimize the voltage difference between inputs." Armed with this heuristic, you can manipulate the gain of the circuit to suit your needs. You can also use it to understand the operation of any operational amplifier.

Let's start our analysis by defining the voltage gain of the generic AC amp in Figure 13-7 as the ratio of output voltage to input voltage. Note that the input voltage is applied to the noninverting (+) input of the operational amplifier. Mathematically, we have this:

$$\text{Gain} = V_{out}/V_{in}(+)$$

Now assume we apply 1.0V to the noninverting (+) input of the operational amplifier. Based on the heuristic, we know that the output feedback will shift such that 1.0V appears at the inverting input (–) of the operational amplifier. Given the circuit configuration, the only way for 1.0V to appear at the inverting input is for the voltage divider formed by R1 and R2 to deliver 1.0V.

Recall that in working with two resistors in series, the voltage drop across a given resistor is proportional to the ratio of resistance contributed by that resistor to the total resistance. For example, consider R1 and R2 in Figure 13-7. It should be intuitive that when R1 = R2, the AC output voltage of the operational amplifier is equally divided between R1 and R2. Similarly, as the value of R1 approaches zero, the voltage drop across R2 approaches the AC output voltage of the operational amplifier. Mathematically, this can be expressed as follows:

$$VR1 = V_{out} \times R1/(R1 + R2)$$

$$VR2 = V_{out} \times R2/(R1 + R2)$$

Following these formulas, given R1 = 25KΩ and R2 = 75KΩ, for a total of 100KΩ, 25 percent of the AC output of the operational amplifier appears across R1 and 75 percent across R2.

Since the voltage drop across R2 is applied to the inverting (–) input of the operational amplifier, we now have this:

$$VR2 = V_{in}(-) = V_{out} \times R2 / (R1 + R2)$$

From our heuristic, we can assume that the values at the inverting (–) and noninverting (+) inputs are equal:

$$V_{in}(+) = V_{in}(-)$$

Now we can express the gain as a function of the output voltage:

$$Gain = V_{out}/[V_{out} \times R2/(R1 + R2)]$$

Simplifying, we have this:

$$Gain = 1 + R1/R2$$

The gain of the generic AC amplifier circuit is defined by the ratio of R1 to R2. Now assume that we need an amplifier that delivers a gain of 10 and that R1 = 1MΩ. We can calculate the value of R2 as follows:

$$Gain = 10 = 1 + R1/R2$$

$$10 = 1 + 1{,}000{,}000/R2$$

$$9 = 1{,}000{,}000/R2$$

$$R2 = 110{,}000\Omega$$

Let's work backward to check our work. Given a 1.0V input to the AC amplifier, at a gain of 10, the output voltage should be 10V. And, as we calculated, 10V across the R1–R2 voltage divider will deliver 1/9th of the output voltage, or 1.0V, to the inverting (–) input of the operational amplifier.

Let's consider the power requirements. The 741, like many operational amplifiers, is designed to operate as a dual-supply device, with positive, negative, and ground. However, providing users with both a positive and negative supply voltage with two batteries and a complicated AC power supply is expensive.

An economical workaround, illustrated in Figure 13-7, is to create a reference voltage from a single-sided supply (+V) and to ground the negative voltage supply lead. In this example, R4 and R5, of equal value, create a voltage divider that supplies a bias halfway between ground and +V, or about 4.5VDC with a 9V supply. Notice how capacitors are used to isolate the inputs and outputs of the operational amplifier from the DC supply voltage and ground.

Because of the high input impedance of the 741 operational amplifier, there must be a finite return path to ground for the input current. In our example circuit, R5, which is typically 1M, serves this function.

Capacitors

The use of capacitors in this pedal is typical of audio circuitry. A 100µf at 16V electrolytic capacitor is used to filter the 9VDC supply. Two 1µf at 35V tantalum capacitors and three Mylar film capacitors, from 0.001 to 0.047µf at 50V, are used for signal handling and bypassing. As shown in Figure 13-8, the 1µf tear-shaped tantalum capacitor is polarized. The capacitance value (simply 1 for 1µf), polarity, and voltage rating are stamped plainly on the capacitor body.

FIGURE 13-8 Tantalum (left) and Mylar capacitors

The Mylar film capacitors, which are dipped in a hard epoxy coating for moisture resistance, are nonpolarized. The code on a Mylar film capacitor indicates capacitance and multiplier. For example, the Mylar capacitor in Figure 13-8, marked *473*, has a capacitance of 47 and a multiplier of 3, or 0.047µf. Similarly, the 0.01µf Mylar capacitor in the pedal is marked *103*, and the 0.001µf capacitor is marked *102*. See Appendix A to review capacitance codes. It's important for you to recognize capacitor values, because much of modding pedals involves replacing capacitors with those of different types and values to achieve different tones.

Connectors

The major connectors in this pedal are for power and audio. The power jack is notable in that it accepts a 2.1mm, 9VDC plug (positive ground) that is standard throughout the pedal industry. If you try to power this or any other effects pedals with a standard negative ground wall wart, you risk frying your pedal.

The mono audio input jack is designed to disconnect power from the circuitry when the plug is removed. If you're debugging a pedal, make sure that you insert a plug in the input jack or you won't be able to activate the signal processing circuit.

This battery-saving feature sounds good in theory, but few musicians take the time to unplug their effects pedals after each practice. Unfortunately, the pedal on–off button is often not a true power on–off switch. For example, with a plug inserted in the input jack of this pedal, the current draw is 0.4mA with the on–off button in the off position. Removing the plug from the input reduces the current draw to zero.

How It Works

This pedal is fundamentally a high-input impedance, symmetrical signal clipper. The audio signal from a guitar is sent to the noninverting input of a 741 operational amplifier configured as an AC amplifier, as shown in the simplified schematic of Figure 13-9. You should recognize most of the major components from the earlier discussion of a generic AC operational amplifier circuit.

Symmetrical signal clipping is the same type of distortion associated with vacuum tubes operated near saturation. By creating symmetrical distortion, this pedal enables a guitar player to achieve the tone characteristic of a powerful tube amp run at full power while using a low-power practice amp.

Gain

As discussed earlier, the gain of the operational amplifier is varied by changing the ratio of R1, a 1MΩ resistor, to R2, the 470KΩ potentiometer. If R2 is set to 470KΩ, the gain of the circuit is as follows:

$$\text{Gain} = 1 + R1/R2 = 1 + 1{,}000{,}000/470{,}000 = 3.1$$

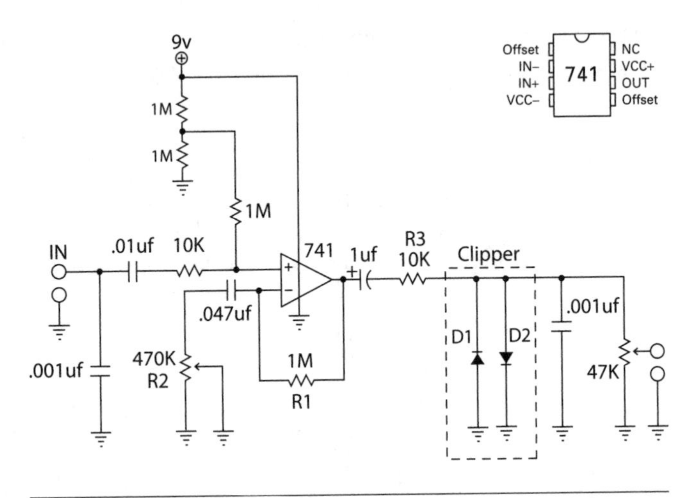

FIGURE 13-9 Simplified schematic of the pedal

With R2 set to 47KΩ, the gain is this:

$$\text{Gain} = 1 + R1/R2 = 1 + 1{,}000{,}000/47{,}000 = 21.3$$

The amplified signal is AC-coupled to a clipper circuit composed of two germanium diodes wired in parallel, anode-to-cathode. The output level is adjusted by varying the tap position of the 47KΩ pot. Because of the clipping effect of the diodes, the theoretical maximum output level of the pedal is ±0.3V, regardless of the gain or level settings.

Clipping

Clipping is performed by two germanium diodes, D1 and D2, protected by a 10KΩ current limiting resistor, R3. The degree of clipping depends on the amplitude of the signal presented to the two diodes, which depends on the setting of the 470KΩ potentiometer. For example, consider Figure 13-10, in which the input signal is approximately ±0.33V. The sinusoidal waveform is minimally clipped at the positive and negative peaks.

Now consider the waveform in Figure 13-11, in which the amplitude of the signal applied across the diodes is approximately ±0.6V. The resulting clipped waveform resembles a square wave, and the sound is harsher than that of the minimally clipped waveform of Figure 13-10.

Figure 13-12 shows an oscilloscope tracing of the pedal input and output signals with the controls set for maximum distortion. I used a BK Precision 4011 function generator to create the 400Hz sine wave input, captured the input and output signal data with a Tektronix TDS2022 100MHz oscilloscope, and plotted the data with a spreadsheet application. Note this figure's resemblance to Figure 13-11.

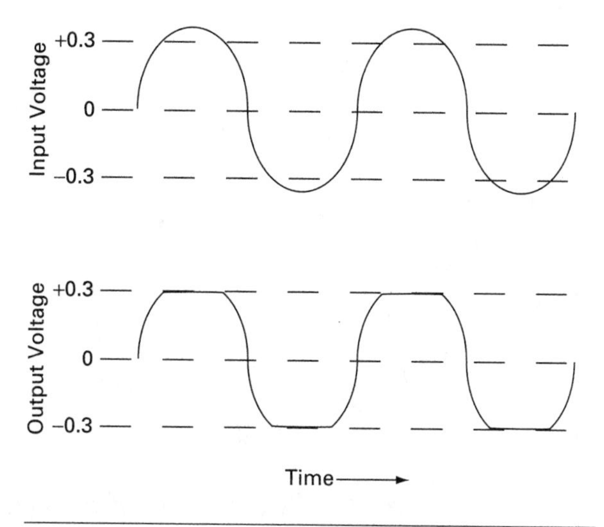

FIGURE 13-10 Minimal clipping

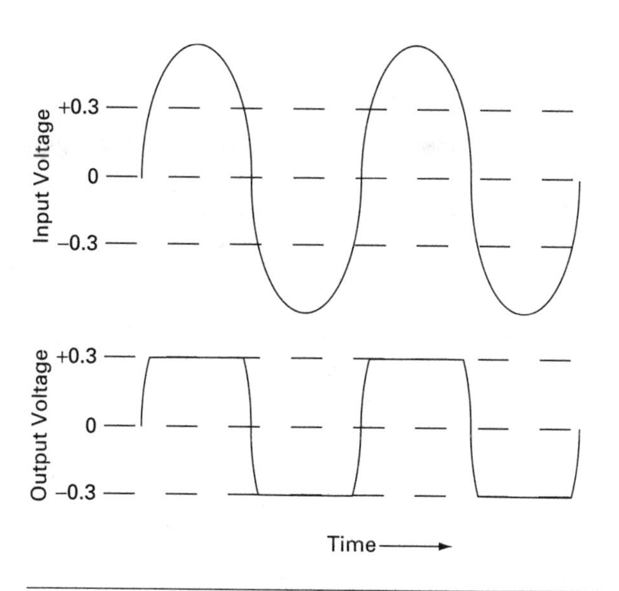

FIGURE 13-11 Pronounced clipping

From examining the figures, you should see that distortion is a function of amplitude of the waveform output from the operational amplifier, relative to the forward conduction voltage of the diode clipper circuit. The lower the resistance setting of the 470KΩ potentiometer, the greater the amplification provided by the operational amplifier, and the greater the distortion. It's also worth noting that the output of the pedal is a little greater than the theoretical limit of ±0.3V.

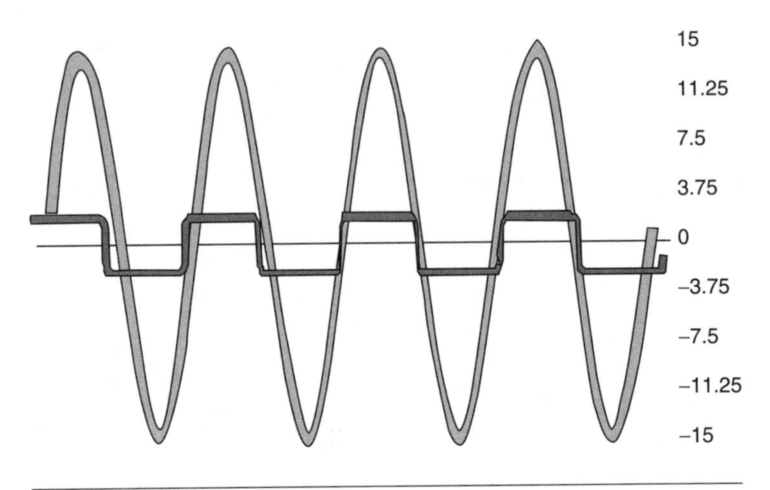

FIGURE 13-12 Oscilloscope trace of clipping

Impedance Matching

The operational amplifier serves as not only a variable gain amplifier to provide varying degrees of clipping, but it also provides a high impedance input for the guitar signal. An electric guitar with standard magnetic pickups is a low-power device designed to work with a high impedance load—on the order of 1MΩ. A lower impedance load will markedly diminish the guitar signal strength and adversely affect tone—notably, the loss of higher frequency signals. Feeding the guitar output directly to a diode clipper would be equivalent to shorting the guitar output through the relatively low-resistance current-limiting resistor.

Mods

There's a healthy economy in pedal modding. For $20, you can purchase a pedal-specific kit with a couple capacitors, diodes, some wire, and a one-page guide of where to install the components. It's common to replace silicon with germanium diodes, a Mylar capacitor with a polypropylene capacitor of higher or lower value, and to replace resistors with wire or resistors of higher value. The value of a kit is information on which capacitors and diodes to replace to achieve a given tone on a specific pedal.

Some generic mods will work with a range of pedals. Following is a list of generic mods that will work for distortion pedals based on a diode clipper.

Change Diode Configuration and Type

Consider different types and configurations of clipping diodes, such as those explained here and shown in Figure 13-13. Each of the clipping circuit configuration circuits depicted here provides a different output tone. The best tone—and therefore configuration—depends on the type of music you want to play and your personal preference. Most guitarists associate the tone produced by germanium diodes with that of vintage effects pedals.

Here's a key to the configurations shown in Figure 13-13:

A. The default version of the pedal, with two germanium diodes, which produces a symmetrical clipping pattern at ±0.3V.

B. A silicon diode paired with a germanium, resulting in an asymmetric clipping at 0.3V on one side of zero and 0.7V on the other.

C. A pair of silicon diodes, resulting in symmetrical clipping at ±0.7V.

D. A silicon diode paired with two germanium diodes in series, resulting in an asymmetrical clipping at 0.7 and 0.6V.

E. A germanium diode paired with two germanium diodes in series, with an asymmetrical clipping at 0.3 and 0.6V.

F. A silicon LED paired with a pair of germanium diodes in series, resulting in an asymmetrical clipping of 1.2 and 0.6V. LEDs conduct when forward-biased at 1.2 to 2.4V, depending on the LED design and composition.

In addition to diodes, you can try transistors and other components with nonlinear junction properties.

Tone Switch

Consider installing a multiposition rotary switch to select one of the diode configurations defined in Figure 13-13. Because of limited space, you'll probably have to mount the switch on the upper side panel of the pedal.

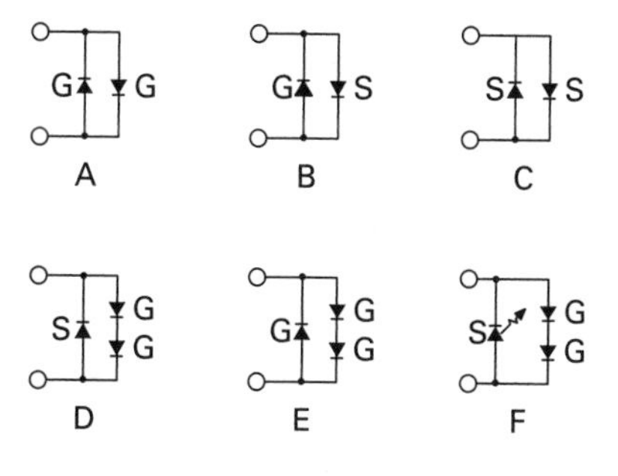

FIGURE 13-13 Clipping circuit configurations

Add True Bypass Switching

The pedal is never completely out of the signal line, even when the on–off button is off. There is a 0.001μf capacitor across the input connector and less than 2MΩ to ground through the bias voltage divider. The on–off button should disconnect these passive elements in the off position to avoid influencing the audio signal.

Try Different Operational Amplifiers

Operational amplifiers of the same type from different manufacturers and op amps of different types have different sound characteristics. To experience the range of possible sounds from your pedal, consider replacing the 741 with an 8-pin DIP (dual inline package) socket so that you can easily change and evaluate different op amps. Check the Web to find modern, high-performance and low-noise operational amplifiers wired to 8-pin DIP carriers compatible with the pin-outs of a 741 operational amplifier.

Chapter 14

Vacuum Tube Guitar Amplifier

In this chapter, I'll tear down the Fender Champion 600, or Champ—a compact 5W combo guitar amplifier that features two vacuum tubes and a 6-inch speaker, as shown in Figure 14-1. The modern Champ is a Chinese reproduction of a small practice amp produced in the United States in the 1950s. It's a favorite of the modding community because of its simple design, affordability, and ready availability of third-party kits and components. Musicians are drawn to the amp because it develops full distortion at only a few watts—a must for bedroom jam sessions when nonmusician parents or roommates don't want to be bothered by the noise.

The Champ retails for about $250, but I've picked up several used amps online for about half that—and spent the difference on various mods. I'll walk you through the circuitry and suggest a few mods that can be applied to most tube-based amps.

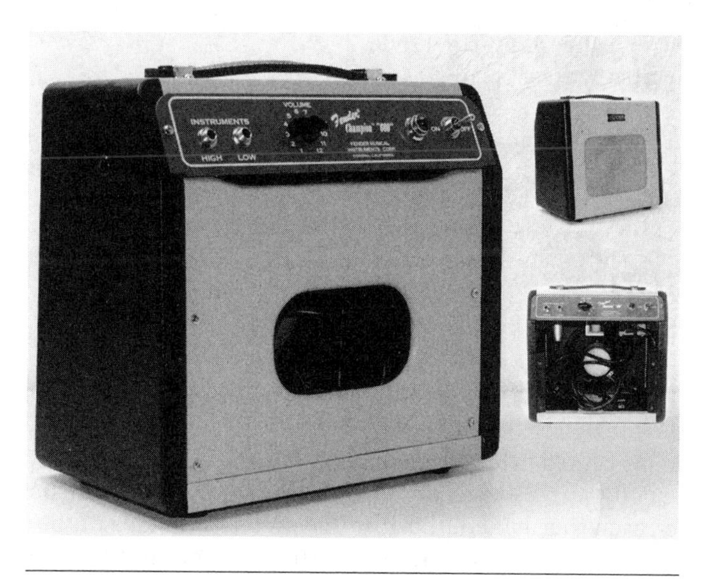

FIGURE 14-1 The Fender Champion 600

Highlights

In this teardown, we'll encounter a high-voltage linear power supply, vacuum tubes, and both power and output transformers. We'll also explore the operating characteristics of Class A amplifiers. You should pay particular attention to the following:

- How shielding is used to minimize noise and EMI
- Provisions for handling heat
- The relatively sparse, open circuit board design
- The use of high-power and high-voltage components
- How the design differs from conventional solid-state devices, in terms of component density, circuit complexity, circuit density, and component count

Specifications

In contrast with the hi-fi industry, Fender and other guitar amplifier suppliers don't provide the typical specifications of their amps, such as frequency response, total harmonic distortion (THD), or noise level. The specifications offered by Fender for the Champ include the following:

- **Dimensions:** 11 × 12 × 7 1/2 inches (HWD)
- **Weight:** 15 pounds
- **Output power:** 5W
- **Speaker:** 4Ω, 6-inch ceramic magnet
- **Preamp tube:** 12AX7A dual triode
- **Power tube:** 6V6GT pentode
- **Inputs:** two

Power input requirements are not specified, but based on my measurements, the amp draws 380mA at 120VAC, or about 40W, regardless of input.

Operation

Operating the Champ involves little more than allowing the tubes to stabilize, plugging in your electric guitar, and turning the volume control clockwise until you like what you hear. Using Figure 14-1 as a guide, on the faceplate, from right to left, is an on–off switch, an LED power indicator, a potentiometer for setting volume and tone, a low-impedance microphone jack, and a high-impedance electric guitar jack.

I usually warm up my amp for about 15 minutes before using it to give the tubes time to reach thermal equilibrium. To give you an idea of the heat involved, consider that after playing for an hour, the temperature of the steel faceplate on my Champ is about 200°F.

Teardown

You should allocate about 2 hours for this teardown, which is illustrated in Figure 14-2.

If you follow along with your own amp, you could be exposed to potentially lethal voltages. Now is a good time to review the safety tips in the Introduction of this book.

Tools and Instruments

For this teardown, you'll need a Phillips-head screwdriver, preferably with insulated handle; a clip lead with medium to large jaws; and a multimeter. Place a towel, carpet remnant, or foam sheet on your workbench to protect the amp's easily torn blonde vinyl.

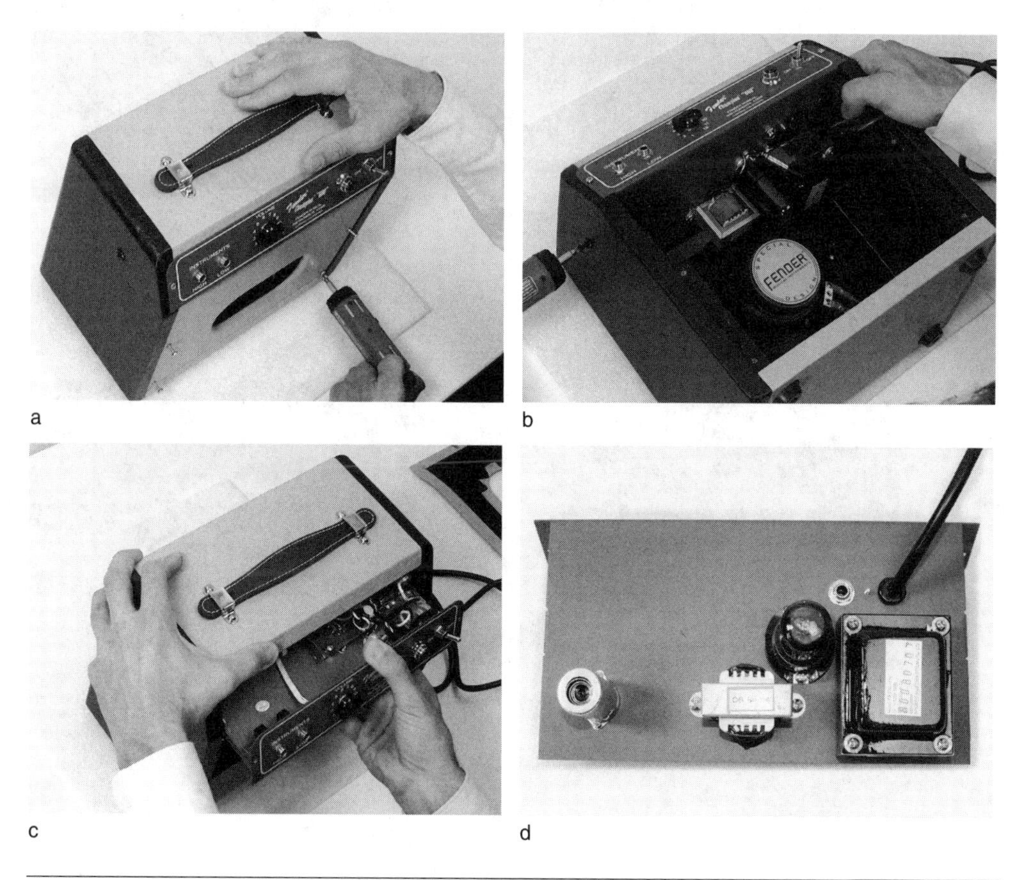

FIGURE 14-2 Teardown sequence

e

f

g

h

i

j

FIGURE 14-2 (*continued*) Teardown sequence

Step by Step

Before embarking on this teardown, take steps to insure your longevity. If you're not experienced working with high-voltage circuits, unplug the amp and set it aside for a day or two to give the high-voltage filter capacitors time to self-discharge. Because the high-voltage filter capacitors lack bleeder resistors, the high voltage will bleed off over hours instead of minutes.

Step 1

Remove the back plate. As depicted in Figure 14-2a, remove the four Phillips-head wood screws that secure the back plate, which is made of thin pressboard, with the vinyl covering stapled from the inner side.

As shown in Figure 14-1, the rolled steel chassis is housed in the upper 2 inches of the cabinet, with the lower part of the cabinet dedicated to housing the speaker. Steel is used instead of aluminum with many tube-type amps because of the mechanical stability it affords.

Step 2

Remove the chassis. Place the amp speaker-side down and remove the two Phillips-head machine screws on either side of the unit, as shown in Figure 14-2b. Next, remove the two screws on either edge of the faceplate. Unplug the speaker jack from the chassis.

With the amp upright, as shown in Figure 14-2c, carefully slide the chassis out of the cabinet. Avoid jarring the fragile glass envelope of the vacuum tube on the underside of the chassis, and don't touch the circuit board.

Step 3

Examine the chassis layout. Carefully flip it over so that the transformers and tubes are facing up, as shown in Figure 14-2d. Notice the large power transformer on the right and the small audio output transformer near the midline of the chassis. The large, exposed 6V6GT power tube is nestled between the two transformers, the audio output jack, and the power cord entrance. The small 12AX7A dual triode, shielded by a steel cover, is to the left.

Step 4

Examine the cabinet. The wood and vinyl cabinet houses a 6-inch speaker (see Figure 14-2e) and a fragile EMF shield composed of thin aluminum foil. The shield extends about 1 inch below the chassis and makes electrical contact with the chassis at the two tabs on either end. The delicate foil, shown in Figure 14-2f, is inevitably damaged as part of the chassis extraction process.

Step 5

Remove the 6V6GT power tube. Use two fingers of one hand to push down on the retaining spring that secures the base of the tube. With your other hand, grip the base collar—not the glass envelope and not the metal tube pins—and pull straight up. Notice the keyed pin at the center of the base, which will help you align the tube's metal pins with the socket when you replace the tube. Remove and set aside the two screws and lock washers that secure the retaining ring.

Step 6

Remove the 12AX7A tube and shield. Push down on the tube shield and twist a few degrees counterclockwise to disengage the two prongs on the shield base. Allow the spring to push up the cover. Grasp the glass envelope of the tube and pull straight back, perpendicular to the chassis, to extract the tube, shown in Figure 14-2g. Because there is no slotted pin in the center of the tube, you'll have to visually align the pins when you reinsert the tube.

Step 7

Examine the circuit boards. Grasp the large power transformer and flip the chassis to expose the two circuit boards, shown in Figure 14-2h. The main board containing the charged high-voltage electrolytic capacitors is secured by six Phillips-head machine screws, and the smaller board is attached to the faceplate by the volume control and input jack hardware. A ribbon cable connects the two boards.

Step 8

Discharge the high-voltage capacitors. With one hand in your pocket, clip one end of a clip lead to one of the side tabs on the chassis. Clip the other end of the clip lead to the metal shaft of an insulated screwdriver. Now carefully touch the tip of the screwdriver to the anode leads of each silicon power diode. Don't be surprised if you see an arc—it's a sign that you've succeeded in at least partially discharging the capacitors. Verify your work with your multimeter, and repeat the process if the circuit isn't fully discharged.

Step 9

Extract the main circuit board, shown in Figure 14-2i. Unplug the ribbon cable connection to the two boards, unplug the power and transformer leads, and remove the six retaining screws. Although the connection points on the board are labeled, make a note of the transformer and power lead connections.

If you didn't remove the retaining ring of the larger vacuum tube, you won't be able to extract the circuit board because the white ceramic tube sockets are soldered to the main board, as shown in Figure 14-2j. Ceramic provides superior insulation and voltage-handling capabilities compared with plastic tube sockets. You'll have a much easier time of examining the components if you unsolder the audio out and LED indicator twisted pairs that tether the board to the chassis.

Layout

The layout of this amp is utter simplicity—one reason why it's the target of so much mod activity. Position the chassis with the on–off switch in the lower-right corner, as shown in Figure 14-3.

Most of the power supply components are on the right side of the main board, immediately behind the DPDT (double-pole, double-throw) on–off switch and LED power indicator. The four silicon power diodes, each with ceramic disc bypass capacitors, and three 22µf at 450VDC electrolytic capacitors are positioned opposite the power transformer. One of the 22µf capacitors serves as a cathode bypass capacitor for the 6V6GT, and the other two are part of the power supply filter. Two fuses, a 1A slow-blow for the 120VAC mains and a 4A slow-blow for the tube filament supply, are nestled in this area as well.

With the exception of a fourth 22µf at 450VDC electrolytic capacitor immediately adjacent to the 12AX7A tube socket, the right side of the board is devoted to signal amplification. Note that the voltage and power ratings of many components are greater than those in a typical solid-state amp. Components are bulky and leaded, typified by the 2W resistors and orange Mylar signal capacitors rated at 400VDC.

The only notable component on the smaller circuit board is the volume control potentiometer. Instead of a common 2W metal potentiometer, the volume control

FIGURE 14-3 Layout

is a compact plastic unit. Although it looks out of place with otherwise vintage components, its silky operation imparts a quality feel to the amp.

Components

The major components in this teardown are the vacuum tubes, transformers, and the elements of the high-voltage power supply.

Vacuum Tubes

The Champ's two vacuum tubes are its most distinguishing features. The small 12AX7A dual-triode amplifier has a characteristic stubby evacuated glass envelope and exposed pins. Each triode, which consists of a heater, cathode, control grid, and plate (the heater and cathode count as one element), can produce a voltage gain of up to 100. The 12AX7A excels in low-level audio applications that require high gain.

The maximum plate dissipation of the 12AX7A, the maximum power the plate element can safely dissipate as heat, is a little more than 1W per triode section, or about 2W for the tube as a whole. Exceeding the maximum plate dissipation because of excessive plate current will literally melt the plate. Furthermore, exceeding the maximum plate voltage of 330VDC will result in internal arcing. In either case, the tube and associated circuitry would be destroyed.

The heaters of the 12AX7A are similar to those of other dual tubes in that they can be connected in serial or in parallel, depending on the available voltage. In this application, the two heaters are connected in parallel and supplied by 6.3VAC. Increasing the heater voltage beyond this will result in overheating and premature failure.

The concentric structure of a triode parallels that of a capacitor. As a consequence, significant capacitance exists between the control grid and plate and can result in unwanted oscillation. This unwanted capacitance can be reduced and the stability of the tube improved by inserting additional grids between the control grid and the plate, as in the tetrode (one additional grid) and pentode (two additional grids) tube designs.

The 6V6GT is a single-pentode power tube consisting of a heater, cathode, control grid, screen grid, suppressor grid, and plate. (Again, for categorization purposes, the heater and cathode count as one element.) Although popular as an output tube for audio amplifiers, the 6V6GT is relatively lightweight as power tubes go. Specifications include an amplification factor of about 10, a plate dissipation of 14W, and a maximum plate voltage of 350VDC. The 6V6GT heater requires 6.3VAC.

The majority of tubes sold in the United States are imported from Russia and China. For example, my Champ's 12AX7A was made by a Russian manufacturer and the 6V6GT is from China.

Transformers

By weight, the iron core power and audio output transformers are the most significant components of the Champ. The power transformer has a 120VAC primary and two secondary windings. The high-voltage winding supplies about 280VAC to the capacitor input power supply, and another secondary winding supplies 6.3VAC to the tube filaments. There are two filaments in the 12AX7A and one in the 6V6GT.

The audio output transformer, unshielded and considerably lighter and smaller than the power transformer, enables the 6V6GT to drive the 4Ω speaker. The transformer isolates the high voltage in the output tube circuit from the output speaker and provides impedance matching. In addition, the audio output transformer enables the high-impedance tube circuit to drive the low-impedance speaker.

Printed Circuit Boards

The main circuit board is notable because there are no foil traces on the top side of the board. In addition, as shown in Figure 14-2j, the foil tracks on the underside are wide and generously spaced to accommodate the relatively high currents and voltages in the amp.

The original Champ, like some modern, boutique guitar amplifiers, sports point-to-point wiring. Components are soldered to posts by their leads, and insulated wire is connected to posts to form circuits. Point-to-point wiring was largely abandoned in favor of the circuit board because it was labor-intensive and didn't lend itself to automated fabrication. Although I've used guitar amps from Carr (www.carramps.com) and Gretsch (http://gretsch.com) with point-to-point wiring, I attribute their wonderful tone to quality components, intelligent circuit design, and sturdy enclosure, rather than the antiquated construction techniques.

Speaker

Externally, the 6-inch Fender speaker is unremarkable. The nominal impedance, essentially the average resistance at range of audio frequencies, is also typical at 4Ω. The DC resistance of my speaker is 3.6Ω, which is normal. However, this speaker, like most guitar amp speakers, differs from most hi-fi speakers in frequency response range and intended operating power. Typical guitar speaker frequency response is 75Hz to 5kHz, which is broader than a standard woofer or midrange speaker.

Another distinguishing features of guitar speakers is extended cone and coil travel. In practical terms, this affects optimum operating power. While a typical stereo speaker rated at, say, 100W, can be driven from 100mW to 100W, the Champ's speaker should be driven at 4–5W so that it can contribute to the highly valued distortion.

Fuses

The Champ's power supply has two fuses: a 1A slow-blow on the primary of the power transformer and a 4A slow-blow in the power transformer 6.3VAC secondary. Because the fuses are slow-blow, occasional transients over the stated amperage won't melt either fuse, but overloads of a few seconds or more will.

Unfortunately, if the Champ suddenly dies, the chassis has to be extracted from the speaker enclosure to examine the fuses. Most modern equipment designs incorporate a fuse holder that's accessible from the outside of the device. Curiously, the original Champ sported an external fuse holder where the on–off switch is today.

Resistors

The plastic potentiometer used to control amp volume is an audio taper potentiometer. As discussed in previous teardowns, the resistance between the middle wiper terminal and a terminal is a logarithmic function of the mechanical position of the wiper.

The fixed resistors are 0.25 and 1W metal oxide film. In addition, the 1W resistors have a flame-resistant, nonflammable coating. Flame-resistant resistors are commonly used in high-voltage circuits where there is potential for shorted tubes and other mishaps that could ignite ordinary components.

Wiring and Cables

The transformers are connected to the main board with PVC-coated stranded wire and clip-on connectors. In addition, the main board is connected to the audio output jack via 18-gauge PVC-insulated stranded wire.

The wires carrying audio are tightly twisted, as are wires from the main board to the LED power indicator. Twisting provides some immunity to EMI and unintended coupling between circuit elements.

LED Power Indicator

The LED power indicator, which is incorporated into a vintage-looking incandescent lens hood, is powered by the 6.3VAC secondary of the power transformer. Moreover, the voltage for the LED indicator is taken after the 4A heater fuse. As a result, if the heater fuse blows, the LED power indicator will not be illuminated, even if the high-voltage circuitry is energized. Bottom line: don't trust the LED power indicator.

Capacitors

The most prominent capacitors in this amp are the four 22µf at 450VDC electrolytic capacitors. Three are used to filter the high-voltage supply. The fourth and the two small axial-lead electrolytic capacitors are used to bias the tubes. The dielectric used in electrolytic capacitors dries out with time, and the elevated temperature of a tube circuit accelerates the process.

Silver mica capacitors are used for high-voltage blocking and signal coupling between amplifier sections, and low-cost ceramic disc capacitors are used for bypassing the power diodes. Both silver mica and ceramic disc capacitors are more stable than electrolytic capacitors, and their operating temperature range is compatible with the elevated temperature of tube circuitry.

Silicon Power Diodes

The amp uses five 1N4006 silicon power diodes, conservatively rated at 800V PIV and 1A average output current. Four are used as a full-wave bridge rectifier in the high-voltage power supply, and one rectifies the 6.3VAC heater voltage to drive the LED power indicator.

The four diodes in the high-voltage bridge are each bypassed with 8200pF ceramic disc capacitors. While the 1N4006 has good inrush current properties, it is susceptible to reverse voltage spikes. The bypass capacitors help protect the diodes against reverse spikes from the power transformer.

EMI Shielding

The aluminum foil EMI shield is at best unimpressive. Electrically, the thin aluminum foil shield should reduce EMI. However, as noted earlier, the foil is fragile and easily damaged.

How It Works

Figure 14-4 shows a simplified schematic of the Champ, including labels for the basic tube elements. A detailed schematic is available on the Fender web site (www.Fender.com).

Starting with the power supply, at the lower-right of the schematic, is a simple, unregulated power supply built around a transformer (T1), a diode bridge, and a capacitor input smoothing filter. The power transformer has two secondary windings, one 280VAC for the high-voltage supply, and one 6.3VAC for the three tube heaters—two in the 12AX7A and one in the 6V6GT—and the power indicator LED. For clarity, the LED circuit, consisting of a diode, series resistor, and LED, is not shown.

The 6.3VAC winding is notable in that it's connected to an artificial center tap composed of a 100Ω resistor on either end of the winding to ground. The benefit of this design over a floating 6.3VAC winding is reduction in hum. A better design is a transformer with a real center tap, but center-tapped transformers cost a few cents more than a pair of resistors.

The 280VAC secondary is connected to a bridge rectifier composed of four silicon power diodes. The resulting full-wave rectified DC is smoothed by a bank of three

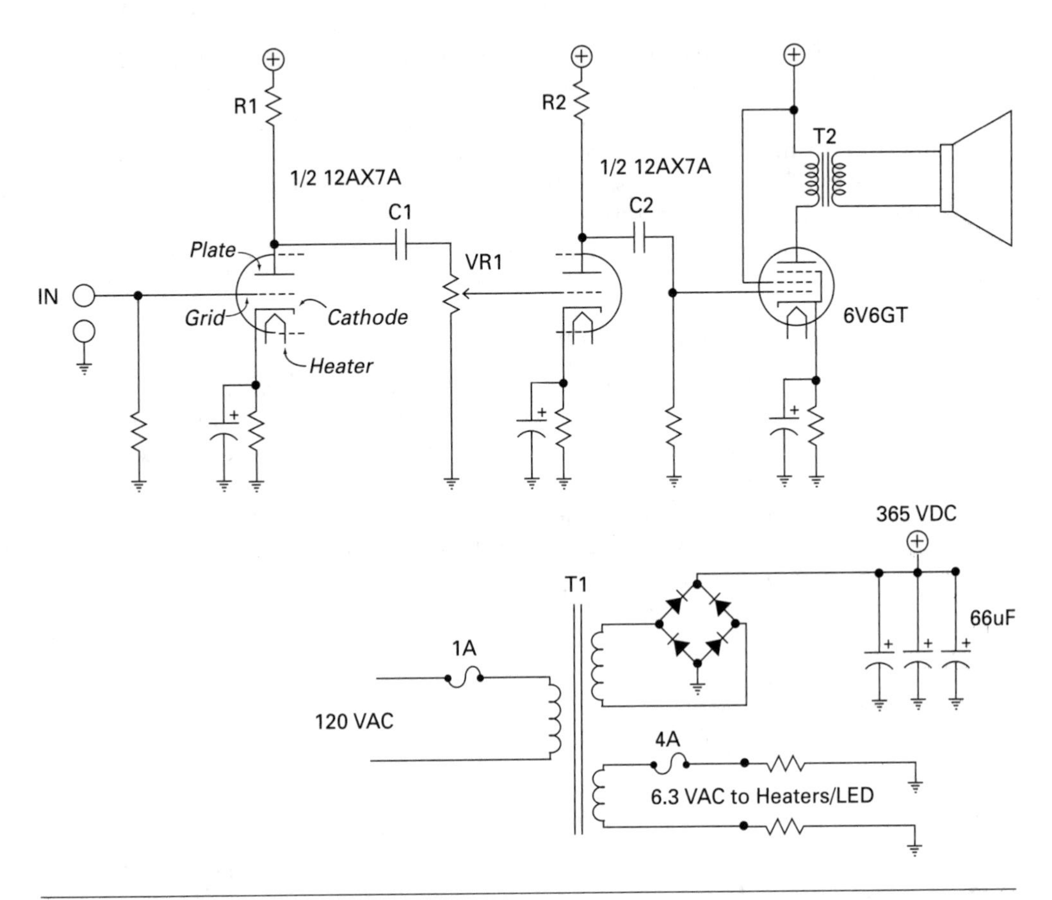

FIGURE 14-4 Simplified schematic

22µf at 450VDC electrolytic capacitors connected in parallel. The capacitor bank, which provides a total of 66µf at 450VDC, boosts the bridge output voltage to about 365VDC, the plate voltage of the 6V6GT output tube. Series resistors (R1 and R1) are used to drop the plate voltage down to about 340VDC for the 12AX7A preamp tube.

The power transformer, which is shielded to reduce hum, is both a step-up and a step-down transformer, supplying both higher and lower voltages than the line voltage applied to the primary winding. Recall that the primary and secondary voltages are related to the turns ratio. Given an input of 120VAC and an output of 280VAC, the primary-to-secondary turns ratio of the power transformer should be 120:280, or 1:2.3, for the high-voltage winding. Similarly, the turns ratio for the primary to 6.3VAC secondary is 120:6.3, or 19:1.

The audio output transformer is a step-down transformer, converting the high voltage, low current of the tube circuit to low voltage, high current for the speaker. While primary and secondary voltages are related by simple turns ratios, primary and secondary impedances are related by turns ratio squared. Given that the typical plate load resistance for a 6V6GT is about 5K and the nominal impedance of Champ's speaker is 4Ω, we can estimate the turns ratio of the audio output transformer. The output-to-speaker impedance ratio is 5000:4, or 1250:1. Taking the square root gives us the turns ratio, or 35:1.

The amplifier circuit proper is built around a two-stage preamplifier using the 12AX7A dual triode. This preamp feeds the 6V6GT power tube, which in turn drives the 4Ω speaker through the audio output transformer. A resistor and capacitor network in the front end of the amp, not shown in the schematic for clarity, provide high- and low-impedance inputs for a guitar and microphone, respectively.

If you're new to tubes, the solid-state equivalent of a triode is an n-channel J-FET (junction field effect transistor). Structurally, the heater element of a 12AX7A triode is surrounded by a cathode, which is in turn surrounded by a porous control grid, which is surrounded by the sheet metal plate. In operation, the heater raises the temperature of the cathode so that it emits electrons. A high positive voltage applied to the plate attracts these electrons. Unimpeded, a plate current flows from cathode to plate. However, a small negative voltage applied to the grid, relative to the cathode, can drastically reduce the flow plate current. In this way, a small varying voltage applied to the grid is amplified in the cathode-plate circuit.

If the control grid is made significantly negative relative to the cathode, electrons streaming from the cathode are repelled and plate current ceases. Conversely, if the grid is made positive relative to the cathode, plate current increases—to a point. A significantly positive grid attracts electrons to itself, resulting in a significant grid current and catastrophic failure of the tube. Establishing the grid voltage relative to the cathode—the tube bias—establishes the operating parameters of the circuit.

Returning to the schematic, the audio signal from the input circuit is applied to the control grid of the first triode. The voltage developed across the plate resistor is coupled through a capacitor (C1) and the volume potentiometer (VR1) to the grid of the second triode. The capacitor blocks the high voltage of the plate circuit while passing the audio signal.

Similarly, the output of the second triode is coupled via C2 to the control grid of the 6V6GT power tube. Variations in the plate current of the 6V6GT are coupled through the audio transformer to the 6-inch speaker. Notice that the first grid after the control grid, the screen grid, is connected to high positive voltage. The grid nearest the plate, the suppressor grid, is connected to the cathode.

A cathode resistor—a resistor in series with the cathode—creates the bias for each tube. The voltage drop across a cathode resistor from plate and grid current makes the cathode positive relative to the control grid. Recall that a negative control grid repels electrons and diminishes plate current. In this way, the amplitude of the negative grid bias affects the amplification factor and the amount of distortion imparted by each tube.

As shown in the schematic, the cathode resistors are bypassed with decoupling capacitors. This decoupling allows for higher amplification because audio-frequency plate current that bypasses the cathode resistor doesn't contribute to a greater voltage drop across the cathode resistor—and a more negative control grid bias voltage.

Without the bypass capacitors, the bias voltage will fluctuate with the input signal. For example, if the input is going positive, the control grid becomes less negative, and plate and therefore cathode current increases. The voltage drop across the cathode resistor makes the control grid more negative, reducing the tube conduction and gain.

The values of the cathode resistors and bypass capacitors are selected so that the plate current is never cut off, regardless of input. This configuration is commonly referred to as a *Class A amplifier.* A characteristic of Class A amplifiers is inefficiency, because the amplifier tube is always conducting, even when there is no input signal. Class A amplifiers are also capable of faithfully reproducing audio with only one output device.

Note the although capacitor coupling is used between tubes, the output of the 6V6GT is coupled to the speaker by the audio output transformer. Transformer coupling from a high-impedance tube to a low-impedance speaker is more efficient than capacitive coupling. Also, recall from our teardown of the solid-state stereo amplifier (Chapter 9) that the low-impedance output of the integrated amplifier chips was connected directly to the speakers.

Mods

Following examples posted on the Web and through my own experimentation, I've made several mods of the Champ to improve the tone, enhance safety, improve usability, and minimize susceptibility to noise. While amplifier specifications such as power and frequency response may be interesting to electronics enthusiasts, guitar players are primarily concerned with tone—a subjective assessment that defies technical definition. Keep this in mind as you consider potential mods for the Champ or another tube guitar amp.

Convenient Fuse Placement

The internal fuses are a pain to check. You can make the mains fuse more convenient to check and change by installing a mains fuse holder and 1A at 120VAC slow-blow in the faceplate, using the space for the low input impedance jack. You'll have to unsolder the unused jack from the small circuit board to make room for the fuse holder. Wire the fuse so that it is on the hot (black) mains lead, between the wall cable and the on–off switch. If the switch shorts, the fuse will blow before the chassis melts. Leave the 4A heater fuse in place.

Improved EMI Shielding

At a minimum, consider reinforcing the area where the chassis tabs make contact with the aluminum foil shield. Use wood screws to attach a 1-inch-wide strip of aluminum—*not foil*—on either side of the inner cabinet. Alternatively, replace the entire shield with copper or aluminum sheeting. Another option is to coat the inner surface of the amp with conductive paint sold through luthier supply stores such as Stewart-MacDonald (www.stewmac.com).

Larger Speaker and Cabinet

The stock speaker, while good at relatively low volumes, bottoms out at maximum volume. It's a simple matter to replace the stock Fender speaker with a 10 or 15W speaker from Celestion, Jensen, or Eminence. If you want to move to an 8-inch or 10-inch speaker, a number of vendors offer cabinet upgrades that accept the Champ's metal chassis.

Improved Thermal Management

The sealed chassis mounted at the top of the cabinet creates an excellent environment for causing early failure of electrolytic capacitors. The simplest thermal management mod is to remove the back plate, at the possible expense of tone. Vacuum tubes are designed to run hot, especially if you want to hear their characteristic distortion.

Another option that doesn't affect tube temperature or performance is to perforate the cabinet top. This mod will definitely decrease resale value, however.

Cleaner, Safer Power

Clean up the power by adding a commercial EMI filter module where the power enters the chassis. An EMI filter is mandatory if you plan to operate the amp around fluorescent lights, cell phones, and computer equipment.

Consider replacing the electrolytic capacitors with a higher ripple current variety. However, avoid the temptation to increase the capacitance as well. If you do, you could increase the voltage to beyond the maximum for the tubes and components. A higher capacitance also places a greater surge load on the diodes, since an uncharged capacitor momentarily looks like a short-circuit to the diode bridge.

Another option is to transform the unregulated, capacitor input supply to a regulated supply by inserting a choke between the bridge rectifier output and capacitor bank. The downside is that a choke-input filter will drop the DC output voltage to about 90 percent of the output from the bridge. A new power transformer with a higher output voltage is a workaround, but an expensive one. Curiously, the original Champ featured a choke-input power supply. I suspect the expensive choke was dropped from the current design as a cost-saving measure.

Finally, add a bleeder resistor across the capacitor bank to bleed off the potentially lethal high voltages after you turn off the amp. A 250K, 1W flame-proof resistor across the high-voltage capacitors should be sufficient.

Gain Control

As discussed earlier, the 12AX7A and 6V6GT operate with fixed bias, by virtue of fixed cathode resistors and bypass capacitors. If you want to adjust the gain of the amp on the fly, replace the fixed cathode resistor of the 6V6GT with a potentiometer. If you make this modification, be certain to stay within the plate dissipation rating of the 6V6GT; otherwise you'll shorten the life of the tube.

The challenge with adding a gain control is that you need a convenient place to mount it. Unless you want to drill a hole in the cabinet, cabinet back, or faceplate, you'll have to remove either the LED power indicator or one of the two input jacks temporarily.

Change the Cathode Capacitor

Changing the value of the 6V6GT cathode bypass capacitor can significantly alter the tone of the amp. Several mod kits increase the value of the bypass capacitor for thicker tone. However, you can experiment with higher and lower values until you find a tone to your liking.

More Efficient Audio Output Transformer

The stock audio output transformer is an inexpensive model from China. A better transformer, either of conventional iron-laminate design or toroidal, could provide higher efficiency. Bigger isn't necessarily better, but all else being equal, a larger audio output transformer will handle more current before saturating.

Permanent Speaker Connection

Minimize the resistance and power loss of the output circuit by wiring the speaker directly to the output of the audio transformer. Consider soldering the leads to the speaker terminals to provide a secure, low-resistance connection.

Higher Quality Tubes

The stock tubes used in many amps don't provide the best possible tone. Although tone preferences vary from person to person, some tubes are know for their superior tone. As noted earlier, the stock Champ comes with relatively inexpensive tubes of Russian and Chinese manufacture. Several companies import tubes in bulk and then test for gain, optimum bias settings, and other parameters, and they resell the tubes under their value-add label. Groove Tubes (www.groovetubes.com), which is owned by Fender, is one such domestic tube reseller. GT tubes are more expensive than their untested counterparts, but you have a better idea of what you're getting.

Try an Upgrade Kit

The easiest way to mod the Champ is to purchase a commercial upgrade kit that addresses a variety of issues with proven mods. I've had success with a combination of a commercial upgrade kit from Mercury Magnetics (www.mercurymagnetics.com) and a new set of tubes from Groove Tubes.

The Mercury Magnetics upgrade includes a power transformer, audio output transformer, choke for the power supply, a handful of components for modifying tube bias levels, and a 15W, 6-inch speaker, shown on the right of Figure 14-5a. The original components appear on the left of the figure. The upgraded amplifier chassis—about 7 pounds heavier than the original—is shown in Figure 14-5b.

The mod kit from Mercury Magnetics is one of dozens available through and reviewed on the Web, and you can find other sources of magnetics for tube amps. For example, Hammond Manufacturing (www.hammondmfg.com) offers an extensive line of transformers and chokes. The benefits of working with a commercial upgrade kit include consistent, known results. You have to determine whether those results are to your liking and worth the expense.

If you make any modifications to the circuitry, use a fused Variac or other brand of variable transformer to increase the AC line voltage gradually while you monitor the input current. I use a Variac with a built-in voltmeter and a digital Kill A Watt EZ (www.p3international.com) to monitor current, but a Variac with a built-in current monitor is best if you can afford one.

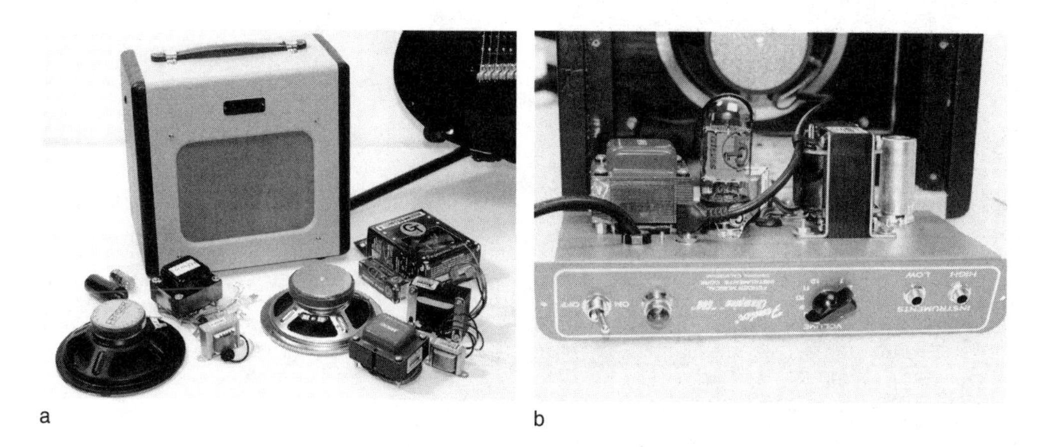

FIGURE 14-5 Mercury Magnetics upgrade

PART IV

Appendixes

Appendix A

Component Markings

To fully understand the function of an electronic device, you'll need to identify the value of the components used in the underlying circuitry. Fortunately, most component manufacturers use standard component markings, which are defined in the following tables.

Numeric Notation

Standard numeric notation is listed in Table A-1. Note that alphabetic notation may be used as a decimal point indicator with SMT components.

TABLE A-1 Numeric Notation

Notation	Multiplier	Example	Value
p – pico	$10^{-12}=0.000000000001$	2.3pF	$2.3 \times 10^{-12}F = 0.0000000000023F$
n – nano	$10^{-9}=0.000000001$	14nH	$14 \times 10^{-9}H = 0.0000000014H$
μ – micro	$10^{-6}=0.000001$	0.01μf	$0.01 \times 10^{-6}F = 0.0000001F$
m – milli	$10^{-3}=0.001$	3mW	$3 \times 10^{-3}W = 0.003W$
K – kilo	$10^{3}=1000$	23KW	$23 \times 10^{3}W = 23{,}000W$
		1K2	$1.2 \times 10^{3}\Omega = 1200\Omega$
M – mega	$10^{6}=1{,}000{,}000$	5MΩ	$5 \times 10^{6}\Omega = 5{,}000{,}000\Omega$
		M47	$0.47 \times 10^{6}\Omega = 470{,}000\Omega$

Leaded Components

With the exception of some capacitors, MOVs, and unmarked glass diodes, most leaded components conform to standard marking systems. The most common standards are reproduced here.

Capacitors

Capacitor marking systems vary by capacitor type. Ceramic disc and most film capacitors are marked in pF using the system in Table A-2. In addition to tolerance markings, film capacitors commonly have a voltage marking. Once you have the system down for a specific vendor or type of capacitor, the most difficult step is converting the value into a range you're familiar with and that's listed in catalogs.

TABLE A-2 Ceramic Disc and Film Capacitor Marking Examples

Marking	Value
101	10×10^1 = 100pF = 0.1nF = 0.0001µf
102	10×10^2 = 1000pF = 1nF = 0.001µf
103	10×10^3 = 1000pF = 10nF = 0.01µf
J	±5%
K	±10%

Electrolytic capacitors, because of their large size relative to other capacitors, have markings for polarity, capacitance, and working voltage. There's generally no guesswork, as the values and units are listed plainly on the component. Where the difficulty arises is deciphering the codes that indicate, for example, whether the capacitor is designed for high temperature, high ripple current, low volume, high stability, and so on. Unfortunately, these codes are vendor-specific. For example, Panasonic Type A, Series EE, is designed for high ripple applications, whereas Type A, Series FM, is a low impedance design.

Resistors

Table A-3 lists the standard resistor and inductor color-code system. Standard value resistors use three bands, the first two for value and the third for the multiplier. A fourth band displays tolerance. Precision resistors may feature a third value band, for a total of five bands.

TABLE A-3 Resistor Color Codes

Color	Value	Multiplier	Tolerance
Black	0	$10^0=1$	
Brown	1	$10^1=10$	±1%
Red	2	$10^2=100$	±2%
Orange	3	$10^3=1K$	
Yellow	4	$10^4=10K$	
Green	5		
Blue	6		
Violet	7		
Grey	8		
White	9		
<none>			±20%
Silver		$10^{-1}=0.1$	±10%
Gold		$10^{-2}=0.01$	±5%

Table A-4 shows examples of color bands applied to standard four-band and precision five-band resistors.

TABLE A-4 Resistor Color-Code Examples

Band 1	Band 2	Band 3	Band 4	Band 5	Value (Ohms)
Black	Red	Red	Silver		1200±10%
Brown	Blue	Orange	Gold		16K±5%
Green	Green	Yellow			550K±20%
Orange	Red	Red	Gold	Red	32.2±2%
Yellow	Grey	Blue	Brown	Brown	4860±1%

Inductors

Leaded inductors use the same basic color-code scheme for value and multiplier as defined in Table A-3. Table A-5 lists several examples of the color-band marking scheme applied to leaded inductors. Note that the value is in µH.

TABLE A-5 Inductor Color-Code Examples

Band 1	Band 2	Band 3	Band 4	Value (μH)
Black	Red	Red	Silver	1200 ± 10%
Red	Violet	Brown	Gold	270 ± 5%
Green	Green	Black		55 ± 20%
Orange	Red	Red	Gold	3200 ± 5%
Yellow	Grey	Brown	Silver	480 ± 10%

SMT Components

SMT components use characters instead of the color bands or dots to indicate value and operating parameters. However, many of the latest generation of SMT components are too small to mark. You'll have to use a capacitance meter to determine the value of those tiny beige ceramic capacitors, for example.

SMT Capacitors

SMT tantalum capacitors use a standard three-digit and one-letter notation, in which the first two digits indicate the value and the third digit indicates the multiplier, and the letter indicates voltage rating. The anode is marked with a polarity band. Depending on the space available on the face of the capacitor, it may also have a batch ID, manufacturer's logo, and date of manufacture code. See Table A-6 for voltage markings.

TABLE A-6 SMT Tantalum Capacitor Voltage Markings

Code	Voltage	Code	Voltage
X	1.8	C	16
E	2.5	D	20
G	4	E	25
J	6.3	V	35
A	10	T	60

The effect of the multiplier on the value represented by the first two digits of a tantalum capacitor can be determined from this formula:

$$\text{Multiplier} = 10^{n-6}$$

where n is the value of the third digit. See Table A-7 for SMT tantalum capacitor marking examples.

TABLE A-7 SMT Tantalum Capacitor Marking Examples

Marking	Value
475J	$47 \times 10^{5-6} = 47 \times 10^{-1} = 4.7$µf at 6.3VDC
476A	$47 \times 10^{6-6} = 47 \times 10^{0} = 47$µf at 10VDC
156V	$15 \times 10^{6-6} = 15 \times 10^{0} = 15$µf at 35VDC
106G	$10 \times 10^{6-6} = 10 \times 10^{0} = 10$µf at 4VDC
227T	$22 \times 10^{7-6} = 22 \times 10^{1} = 220$µf at 50VDC

SMT electrolytic capacitors are marked with a straightforward notation that indicates capacitance in µf and working voltage, as in Table A-8.

TABLE A-8 SMT Electrolytic Capacitor Marking Examples

Marking	Value
1/16	1µf at 16VDC
22/25	22µf at 25VDC
100/25	100µf at 25VDC

SMT Resistors

Standard-tolerance SMT resistors are marked with a three-digit code. The first two digits are the value and the third digit is the multiplier. The multiplier is simply 10^n, where n is the value of the third digit. Resistance less than 10Ω is marked with *R* to indicate the position of the decimal point. See Table A-9 for examples.

TABLE A-9 SMT Standard-tolerance Resistor Marking Examples

Marking	Value (Ohms)
475	$47 \times 10^5 = 4.7M$
4R6	4.6
150	$15 \times 10^0 = 15$
151	$15 \times 10^1 = 150$

Precision resistors are marked with a four-digit code, in which the first three digits represent the most significant digits of the component value and the fourth is the multiplier. See Table A-10 for examples. Note that simply because a value is possible doesn't mean that it's available. Manufacturers produce standard resistor values that are applicable to practical engineering design needs. For example, it would be expensive to manufacture 6.2, 6.3, 6.4, and 6.5K ± 5 percent resistors, in part because their values nearly overlap when you consider the tolerance. Standard values in this range and tolerance are 6.2 and 6.8K. Other values are possible, but nonstandard values are more expensive than standard values.

TABLE A-10 SMT Precision-tolerance Resistor Marking Examples

Marking	Value (Ohms)
4995	$499 \times 10^5 = 49.9M$
0R22	0.22
1000	$100 \times 10^0 = 100$
1501	$150 \times 10^1 = 1.5K$

Appendix B

Resources

Practical Theory

Web sites are a great source for component and circuit theory. My favorite first stop on the Web is Wikipedia (www.wikipedia.org). But logging into the Web can be a hassle in the middle of a teardown. If you're just getting into electronics, you owe it to yourself to invest in a modest library of books or ebooks.

Core Resources

If your bookshelf is bare, I suggest starting with *Getting Started in Electronics*, by Forrest M. Mims III (Master Publishing, 2003). Once you've digested the material, invest in a copy of *The Art of Electronics* by Horowitz and Hill (Cambridge University Press, 1989). If your interests are in communications, power supplies, and RF amplifiers, consider adding any year of the *ARRL Handbook for Radio Communications* to your library. All three books are available through Amazon and Barnes & Noble. Go for the second-hand copies of these classics and save yourself some money.

If you're short on cash or simply prefer a local electronic reference, then you have to download *Lessons in Electric Circuits*, by Tony R. Kuphaldt, at www.openbookproject.net/electricCircuits. This excellent, richly illustrated, and regularly updated resource is available for free download as searchable PDF and HTML files.

Topic-specific Sources

If you've exhausted the core resources and want to delve deeper into a topic, try these books and web sites.

Control Theory

- *Feedback Systems: An Introduction for Scientists and Engineers*, by K.J. Astrom and R.M. Murray (Princeton University Press, 2009). www.cds.caltech.edu/~murray/amwiki
- Parallax, Inc., www.Parallax.com Information on microprocessors and control applications using the BASIC Stamp processor.

Distortion and Music

- Electronic Musician tutorials, www.emusician.com/tutorials

Guitar Effects Pedals

- *Guitar Effects Pedals*, by D. Hunter (Backbeat Books, 2004). While mainly dealing with the history and practical application of effects pedals, Hunter's book offers generic schematics for a range of standard effects.

Guitar Magnetic Pickups

- Fender, www.fender.com
- *The Guitar Pickup Handbook*, by D. Hunter (Backbeat Books, 2008). A good book, written more for the musician than electronics enthusiast.
- "The Secrets of Electric Guitar Pickups," by H. Lemme. http://www.buildyourguitar.com/resources/lemme/index.htm A good overview of the complexities of pickups and system interactions.

PIR Sensors

- Parallax, www.Parallax.com Free, downloadable documentation on an inexpensive passive infrared (PIR) sensor based on the BIS0001.

Semiconductors

- National Semiconductor, www.national.com The web site features an interactive program called WEBENCH Power Designer that enables you to design and simulate circuits.

Tube Amplifiers

- Aiken Amplification, www.aikenamps.com
- *Building Valve Amplifiers*, by Morgan Jones (Newnes, 2004).
- *The Guitar Amp Handbook: Understanding Tube Amplifiers and Getting Great Sounds*, by D. Hunter (Backbeat Books, 2005).

Zebra Elastomeric Connectors

- Fujipoly America, www.fujipoly.com

Component Specifications

The best source for information on components is the component's official datasheet, directly from the manufacturer, a datasheet clearinghouse, or through a parts catalog.

General

Here are my favorite online and print catalog references, in order of preference:

- DigiKey, www.digikey.com My first stop for component specifications. Products are linked to an extensive online database of datasheets.
- Mouser Electronics, www.mouser.com My second stop for general components, and my first for hard-to-find and specialized components.
- DatasheetCatalog.com, www.datasheetcatalog.com Free, extensive database of official datasheets of semiconductors by device type and manufacturer.
- Allied Electronics, www.alliedelec.com Excellent source for capacitor and resistor information.
- Jameco Electronics. www.jameco.com Less extensive than DigiKey or Mouser, but easier to navigate, and the online photos make it easy to identify components.
- Parts Express, www.parts-express.com Like Jameco, easy to navigate and loaded with photos of components. Great for specifications on speakers and audio components.

Component-specific Sources

The companies and web sites listed here include sources of components discussed in the teardowns covered in this book. There are hundreds of additional component companies that can be accessed on the Web.

Capacitors

- Carli Electronics Company, www.carli-cap.com.tw
- NTE Electronics, www.nteinc.com
- RFE International, rfeinc.com
- Vishay, www.vishay.com

Diodes

- Diodes Incorporated, www.diodes.com
- National Semiconductor, www.national.com

Electric Guitar Components

- Allparts, www.allparts.com
- Fender, www.fender.com
- Stewart-MacDonald Instrument Repair Supply, www.stewmac.com

Microcontrollers and Supporting Components

- Atmel, www.atmel.com
- Freescale Semiconductor, www.freescale.com

Operational Amplifiers

- National Semiconductor, www.national.com

Resistors

- Bourns, www.bourns.com Fixed, variable, leaded, SMT, and high power resistors.
- Vishay, www.vishay.com

Speakers

- Celestion, www.celestion.com
- Eminence, www.eminence.com

Temperature Sensors

- Diodes Incorporated, www.diodes.com
- National Semiconductor, www.national.com
- Vishay, www.vishay.com

Transistors

- Diodes Incorporated, www.diodes.com
- Vishay, www.vishay.com

Transformers and Inductors

- Hammond Manufacturing, www.hammondmfg.com
- Vishay, www.vishay.com

Vacuum Tubes

- Tube Zone, www.tubezone.net/tubedata.html An extensive source of freely downloadable PDFs of original vacuum tube manufacturers' datasheets.

Index

D

E

S